Contributions to Environmental Sciences & Innovative Business Technology

Contributions to Environmental Sciences & Innovative Business Technology (CESIBT) is an interdisciplinary series of peer-reviewed books dedicated to addressing emerging research trends relevant to the interplay between Environmental Sciences, Innovation, and Business Technology in their broadest sense. This series constitutes a comprehensive up-to-date interdisciplinary reference that develops integrated concepts for sustainability and discusses the emerging trends and practices that will define the future of these disciplines.

This series publishes the latest developments and research in the various areas of Environmental Sciences, Innovation, and Business Technology, combined with scientific quality and timeliness. It encompasses the theoretical, practical, and methodological aspects of all branches of these scientific disciplines embedded in the fields of Environmental Sciences, Innovation, and Business Technology.

The series also draws on the best research papers from EuroMid Academy of Business and Technology (EMABT) and other international conferences to foster the creation and development of sustainable solutions for local and international organizations worldwide. By including interdisciplinary contributions, this series introduces innovative tools that can best support and shape both the economical and sustainability agenda for the welfare of all countries, through better use of data, a more effective organization, and global, local, and individual work. The series can also present new case studies in real-world settings offering solid examples of recent innovations and business technology with special consideration for resolving environmental issues in different regions of the world.

The series can be beneficial to researchers, instructors, practitioners, consultants, and industrial experts, in addition to governments from around the world. Published in collaboration with EMABT, the Springer CESIBT series will bring together the latest research that addresses key challenges and issues in the domain of Environmental Sciences & Innovative Business Technology for sustainable development.

Jaheer Mukthar K. P. • Edwin Ramirez Asis •
Murugesan T. K. • Justin Nelson Michael
Editors

Business Resilience and Digital Technology in the Post-Pandemic Era

A Global Case

Springer

Editors
Jaheer Mukthar K. P.
Department of Economics
Kristu Jayanti College
Bengaluru, India

Edwin Ramirez Asis
Universidad Senor de Sipan
Chiclayo, Peru

Murugesan T. K.
Department of Commerce (PG)
Kristu Jayanti College
Bengaluru, India

Justin Nelson Michael
School of Management
Kristu Jayanti College
Bengaluru, India

ISSN 2731-8303 ISSN 2731-8311 (electronic)
Contributions to Environmental Sciences & Innovative Business Technology
ISBN 978-3-031-48074-4 ISBN 978-3-031-48075-1 (eBook)
https://doi.org/10.1007/978-3-031-48075-1

This Springer imprint is published by the registered company Springer Nature Switzerland AG
The registered company address is: Gewerbestrasse 11, 6330 Cham, Switzerland

Printed on acid-free paper.

Contents

The Relationship Between Microenterprises' Sense of Entrepreneurial Competence and Their Success in Peru

Giovani Villegas-Ramirez, Luis Angulo-Cabanillas, Juan Villanueva-Calderon, Tatiana Gonzales-Yanac, and Edwin Hernan Ramirez

Abstract The concept of ESE (or "entrepreneurial self-efficacy") has been increasingly central to the study of entrepreneurship in recent years. The study surveyed 382 business owners across four provinces in Peru, with the aim of analyzing and gauging the impact of the five ESE aspects on the commercial success of micro- and small firms. The analysis was performed in SmartPLS 3.3. Business success for micro- and small enterprises was found to be positively impacted by ESE characteristics. It was also determined that the promotion of an inventive atmosphere contributed less to the improvement in company results than the dimension of establishing interactions with ESE investors.

Keywords Business success · Innovative environment · Sense of competence · Work under stress

1 Introduction

Researchers in the field of entrepreneurship have paid a great deal of attention to the concept of entrepreneurial self-efficacy (ESE) since the turn of the century [1]. It's a crucial idea for studies of different kinds of business startups. Although it has been defined somewhat differently by many researchers, its significance in the study of entrepreneurship has been acknowledged by all academics. Numerous studies have

G. Villegas-Ramirez (✉)
Universidad Nacional Santiago Antunez de Mayolo, Huaraz, Peru
e-mail: gvillegasr@unasam.edu.pe

L. Angulo-Cabanillas · T. Gonzales-Yanac · E. H. Ramirez
Universidad Cesar Vallejo, Lima, Peru
e-mail: langulo@ucv.edu.pe; dgonzalesya15@ucvvirtual.edu.pe; eramirezas@ucvvirtual.edu.pe

J. Villanueva-Calderon
Universidad Cesar Vallejo, Chiclayo, Peru

J. M. K. P. et al. (eds.), *Business Resilience and Digital Technology in the Post-Pandemic Era*, Contributions to Environmental Sciences & Innovative Business Technology, https://doi.org/10.1007/978-3-031-48075-1_1

stressed the significance of entrepreneurs' belief in their own abilities in reaching their objectives. As a component of his theory of social learning, [2] coined the term "self-efficacy." A person's confidence in his or her capacity to do a particular set of activities was defined as such. It is the entrepreneur's mental evaluation of his or her own drive, conceptual skills, and action plans to take charge of business needs [3]. Success can be predicted in multiple ways by self-efficacy [4]. Confidence in one's abilities necessary for multiple related domains [5, 6] is the second dimension of competence, as defined by Valencia-Arias and Marulanda-Valencia [7]. The first dimension, "context-specific," refers to an individual's confidence in successfully achieving a specific task. Several scholars took [2] at his word and offered the ESE framework after hearing his ideas. They identified ESE as the confidence to pursue and succeed at entrepreneurial endeavors. Despite a large body of literature on the topic, it is unclear what kind of link ESS has with business success for a micro- or tiny organization [8].

Since self-efficacy has a robust theoretical foundation and excellent academic backing for comprehending prospective Business success [9], our investigation contributes in a number of ways. Based on the definition of SLT provided by Ramirez-Asis et al. [3], SES can aid in "task-specific" results, as stated by Thommandru et al. [2]). Entrepreneurial efficacy attitude (EEA) is characterized by a high degree of confidence in one's own abilities to succeed in entrepreneurial positions and activities [10]. These business owners have the following traits: they aim high but aren't unrealistic, they put in a lot of work on their ventures, and they bounce back quickly from setbacks [11]. Self-confident business owners, according to many studies, are more likely to strive for ambitious expansion plans for their companies and are more likely to persist in their efforts to realize those plans. Persistent efforts like these typically pay off in the end [5]. As a result of the beneficial effects of EEA, their convictions are translated into action, resulting in enhanced business outcomes.

Yating et al. [10] are widely credited with developing the 20-item self-efficacy construct. It included of five factors—New Product Design, Human Resource Development, Enabling Investor Relations, Innovative Environment, and Working Under Stress—and was graded as EEA to help foretell a person's perspective to become an entrepreneur [12]. Constructs were adapted for use in the context of Peruvian microenterprises, and these measures were taken into account based on the work of other researchers who have employed them [13]. As [11] suggest, microenterprises are utilized to analyze the ESE since they play a crucial role in developing nations like Peru, and despite their poor survival rate, their growth rates are sometimes enormous.

Other researchers have cited its difficulty to measure due to various technical problems of accounting practices [14], but revenue, employment, investment, and profitability have been the predominant methods for measuring business success in terms of growth [15]. The importance of measuring business success in small enterprises has also been emphasized by studies [16]. According to [17], firm business performance can be gauged in a number of ways, despite the fact that growth can take several forms. People's confidence in their own abilities to bring about positive change has been linked to higher rates of success in business [10]. Moreover, the expansion of revenue and employment and investment are two particular yet subjective indicators of business performance that have been highlighted in various earlier research [18]. A key differentiator between successful entrepreneurs and those who aren't is, as stated by Higuerey and Armas [16], the pursuit of high entrepreneurial growth.

Researchers have found that business owners have greater confidence in their abilities than the general population. Patent creators who are also involved in the formation of new businesses tend to have a greater sense of self-efficacy than those who are not [10]. Despite its apparent strength, ESE theory is not yet fully developed, and its validity is debated along various dimensions [13]. This research contributes by validating EEA characteristics and establishing a link between EEA and the success of Peruvian microenterprises. It's important to note that there hasn't been a study done of the micro- and small-business environment in Peru before; therefore, this research will help pinpoint the causes and effects of ESE among microenterprises. The primary goal of this analysis is to clarify the function of each ESE element and its impact on the success of micro- and small businesses, which play a significant role in driving economic expansion.

The relevance of microenterprises cannot be overstated in light of the rising unemployment rate in Peru, a developing country experiencing a challenging economic scenario. ESE analysis can be useful for business owners in this field, which is crucial to the nation's economic growth. Enhancing ESE can help business owners run more efficiently, which can boost their company's bottom line in the long run [19]. In addition, this research can help other stakeholders, especially governments, better understand the challenges micro- and small-business owners have while trying to enhance ESE in order to compete in this fast-paced industry. Finally, the government can help microenterprises find and hire people with high EEA across the board to boost their businesses' chances of success.

2 Research and Development Framework

There has been conflicting evidence from recent studies that have evaluated the impact of EEA on the financial success of businesses [20]. Despite the fact that many studies failed to provide evidence for the contradictory links they claimed to find, others have found a positive link between ESE and the financial success of businesses. Tian et al. [21] conducted a meta-analysis of 114 papers to determine the relationship between self-efficacy and business success in the workplace. Several more studies [13] have found a favorable correlation between training and entrepreneurial self-efficacy. Some investigations, such as that conducted by Pallathadka et al. [22], have revealed that ESE has an indirect effect on entrepreneurs and their ventures, despite the fact that many studies have found a positive correlation between ESE and entrepreneurial business success. Because entrepreneurial self-efficacy can be used in many contexts, several research have shown opposite results. Some scholars believe that overconfidence can make business owners conceited or envious, traits that might undermine their effectiveness as leaders and as managers [23]. However, it's possible that the entrepreneurship literature hasn't dealt with this issue yet. Strategic management and behavioral and organizational theory, to name a few, have published on the potential pitfalls of overconfidence in one's own abilities to succeed as an entrepreneur [12]. It is important to note that little research has been done to further examine and assess the correlations between the five essential aspects of entrepreneurial self-efficacy and company business success before discussing the association between the five dimensions [24].

This research seeks to fill this void in the literature by using a novel approach to evaluate the model. There is a correlation between marketing expertise and confidence in one's ability to design and market new products. Self-efficacy in product creation is an important variable that might affect a company's bottom line, according to [25]. Expertise in "strategic product design decision making, sales Business success, and marketing communications" (p. 91) is what [26] mean by this term. They also stress the importance of how an organization's bottom line is affected by the quality of the product design. However, when assessing the characteristics of self-efficacy in the higher-education sector, [24] showed that marketing self-efficacy had a negligible impact on institutional business success. Planning and reducing uncertainty in regard to one's own sense of efficacy is discussed in the context of ESE's human resource management (HRM) in the second dimension [17]. The EEA HRM factor was found to have a statistically significant correlation with franchise business success, according to research by Herath and Mahmood [27], investigating the connection between EEA dimensions and firm business performance, and they discovered that the human resource development dimension of EEA concerned with defining purpose had a positive correlation with company business success. A similar study in Nigeria found no correlation between the ESE's human resource management dimension and business success.

Yating et al. [10] note that investor relations is one of the primary aspects that positively effect a firm's business success, and they introduce investor relations as a dimension of self-efficacy. [27] found a similar favorable and strong association between this aspect of EEA and the success of businesses. For the fourth dimension, creativity and the concept of establishing an inventive environment (BIE) are often discussed in the context of entrepreneurship and are seen as one of the primary "drivers" that aid in the success of entrepreneurs [28]. Experts claim that those with high EEA offer an innovative environment to their firms and set loftier goals because of the close relationship between the two concepts [15]. Innovation, self-efficacy, and business success are all interconnected and mutually affective, as [7] point out. The authors conclude that there is a positive and significant relationship between WUE and franchise business success because of the link between WUE and self-efficacy (as reported by Yating et al. [10] and the belief that entrepreneurs' working under stress habits can lead to better business success. The EEA factor of work under stress was also revealed to be favorably associated with organizational business success in a recent study by Bakar et al. [24]. Initiating investor relations (IIR), the final ESE feature, is connected to better financial analysis and cost control. In terms of how it contributes to the success of a business, it is crucial to increasing

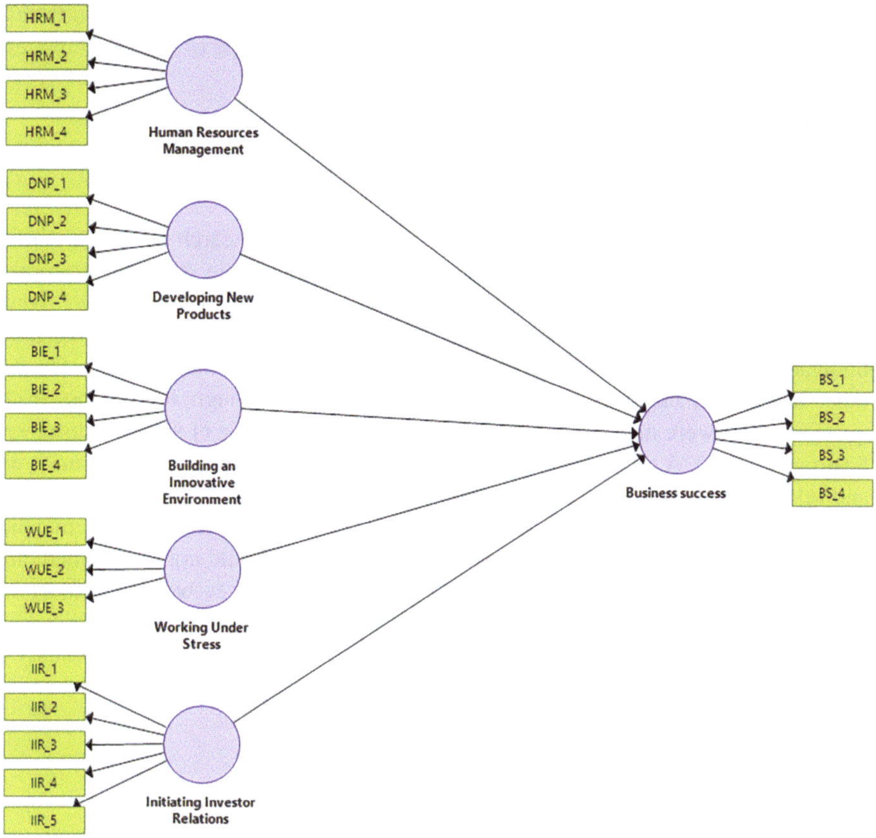

Fig. 1 Research model

productivity inside an organization. Positive and statistically significant links between ESE investor relations and franchise performance are highlighted by Yating et al. [10]. Bakar et al. [24] found a correlation between these two factors and business success (Fig. 1).

According to the research framework, there have been five independent variables, namely, Building an Innovative Environment (BIE), Developing New Products (DNP), Managing Human Resources (HRM), Working Under Stress (WUE), and Initiating Investor Relations (IIR), and a dependent variable named Firm Business Success (DE). In addition, the following five hypotheses have been formulated as part of the literature review section:

H1: Designing new products positively affects the business success of microenterprises

H2: Managing human resources positively affects microenterprises' business success

H3: Initiating relationships with investors positively affects MSE business success

H4: Building an innovative environment positively affects MSE business success
H5: Working under stress positively affects the business success of microenterprises.

3 Methodology

In order to fulfil the purpose of this study, a quantitative research approach with a non-experimental design was used [29]. To assess the variable AEE, the entrepreneurial self-efficacy scale adapted by [10] was used. It has five dimensions: New product design, human resource management, initiating relationships with investors, innovative environment, and working under stress. This scale has 20 items. On the other hand, firm business success has been measured through four items. Both questionnaires were measured using a 5-point Likert-type scale (1 strongly disagree, 5 strongly agree). A total of 392 entrepreneurs in four Peruvian cities were surveyed and selected using stratified random sampling [29]. In this sense, the questionnaires were distributed during the month of November 2022. The COVID-19 pandemic did not affect data collected during this time period; hence, it should not be assumed that this had any bearing on the results. By constructing a measurement model in SmartPLS version 3.3., we were able to examine aspects of construct validity like convergent and discriminant validity in addition to composite reliability. The study's hypotheses and model acceptance rates were put to the test by first applying them to the measurement model.

4 Results

Mean and standard deviation were the first descriptive statistics we calculated when analyzing the data. All values are relatively near to the mean value of 3, as shown by the medians. Working under ESE stress has the highest reported mean value (3.13), indicating that respondents are more likely to take risks, while entrepreneurial business success has the lowest reported mean (0.23), indicating that respondents are less likely to be successful in business. The distributed mean also ranges from 0.65 to 0.93 (Table 1).

Table 1 Descriptive statistics

Latent variables	Mean	Standard deviation
Design of new products DNP	2.99	0.67
Human resources management HRM	3.04	0.78
Initiate investor relations IIR	2.51	0.82
Building an innovative environment BIE	2.84	0.68
Working under stress WUE	3.13	0.65
Business success DE	2.56	0.93

4.1 Model Measurement

Figure 2 shows the results of the factor loadings using the 0.6 threshold provided by Hair et al. (2017), which was employed in the SEM-PLS analysis of the research model using Smart PLS software. All showed convergent validity with values greater than 0.6.

4.2 Convergent Validity

Convergent validity and composite reliability have been calculated and are displayed in Table 2. Composite dependability assesses how well individual components of a measurement system align with one another [30]. The composite reliability threshold is set at 0.70. In addition, "the extent to which items measuring the same construct

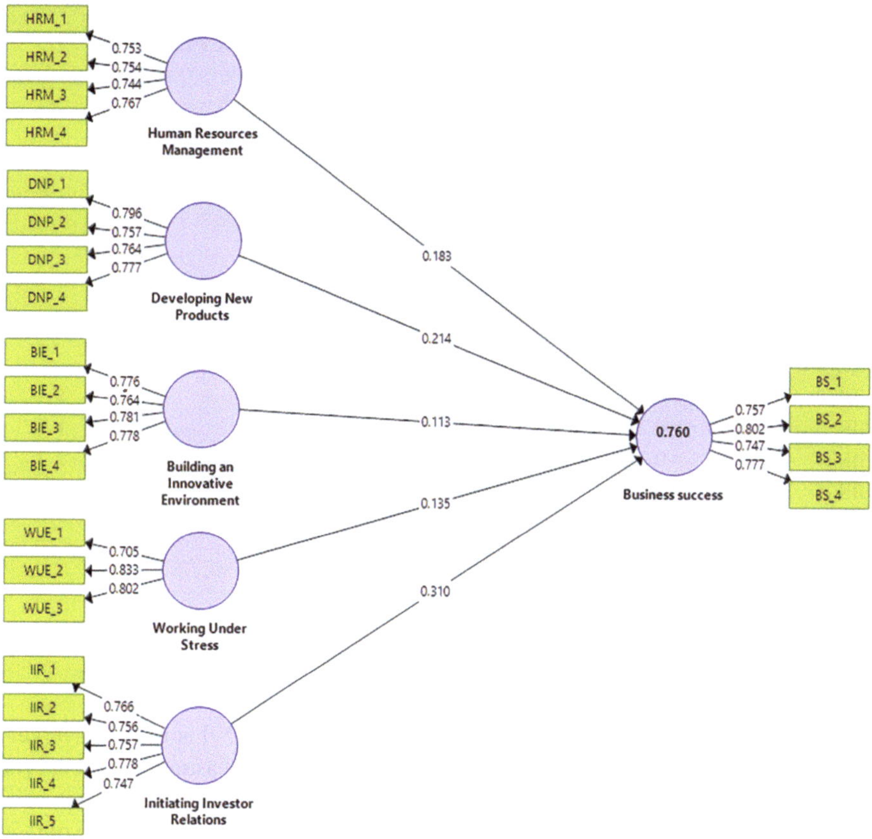

Fig. 2 Research model

Table 2 Reliability and convergent validity

Construct	Standardized Factor Loadings (CFE)	Cronbach's alpha	Composite Reliability (FC)	Average extracted variance (AVE)
Building an innovative environment	BIE	0.779	0.857	0.600
BIE_1	0.776			
BIE_2	0.764			
BIE_3	0.781			
BIE_4	0.778			
Business success	BS	0.773	0.854	0.594
BS_1	0.757			
BS_2	0.802			
BS_3	0.747			
BS_4	0.777			
Developing new products	DNP	0.776	0.856	0.598
DNP_1	0.796			
DNP_2	0.757			
DNP_3	0.764			
DNP_4	0.777			
Human resources management	HRM	0.749	0.841	0.569
HRM_1	0.753			
HRM_2	0.754			
HRM_3	0.744			
HRM_4	0.767			
Initiating investor relations	IIR	0.818	0.873	0.579
IIR_1	0.766			
IIR_2	0.756			
IIR_3	0.757			
IIR_4	0.778			
IIR_5	0.747			
Working under stress	WUE	0.681	0.824	0.611
WUE_1	0.705			
WUE_2	0.833			
WUE_3	0.802			

are correlated with each other and measure the same thing" [29] is what convergent validity means in the case of a test. Convergent validity in a measurement model can be evaluated by meeting three preconditions. The 0.7 threshold of the composite reliability requirement must first be satisfied. The second criterion is an AVE of 0.50

or above, and the third criterion is that the standardized factor loadings be at least 0.60.

Table 2 shows that the dependability of all latent variables as a whole is between 0.84 and 0.873, well above the required 0.70 [30]. The FC has also met the minimum requirements for convergent validity, as we have shown above. Standardized factor loadings likewise were above the 0.70 minimum threshold value, while the AVE, at 0.569–0.611, is above the 0.50 criterion. The model has been shown to have convergent validity, and it can now be used for hypothesis testing.

4.3 Discriminant Validity

According to [30], the concept of discriminant validity refers to the "degree to which items measuring one construct are different from items measuring other constructs." According to [29], discriminant validity can be realized when all correlations between latent components are found to be significantly lower than the square roots of AVE.

According to what is presented in Table 3, the correlations between the two latent variables are less than the square root of the AVE. The numbers in the diagonals, which show that the square root of the mean variance is larger than the other values, are larger than the values in the rows and the columns. This distinction demonstrates, among other things, that the model possesses discriminant validity.

Once the measurement model was achieved, it was verified by bootstrapping at a subsample range of 500. Table 4 shows the testing of the hypotheses; all five tests show that it positively affects. Human Resources Management positively affects business success, H1 ($\beta = 0.183$; $t = 2.948$, $p < 0.01$); similarly, Developing New Products Managing Human Resources positively affects business success, H2 ($\beta = 0.214$; $t = 4.183$; $p < 0.001$); Hypothesis H3, Building an Innovative Environment positively affects entrepreneurial business success is supported by ($\beta = 0.113$; $t = 1.689$; $p < 0.05$); similarly, Working Under Stress positively affects entrepreneurial business success H4: ($\beta = 0.135$; $t = 3.013$, $p < 0.01$); finally, H5, Initiating Investor Relations positively affects entrepreneurial business success due to the values ($\beta = 0.310$; $t = 5.778$; $p < 0.001$).

According to the results, the validity of the model of EEA and microenterprises' business success is demonstrated. Additionally, the R2 measured the effect. The R2

Table 3 Discriminant validity

	BIE	DE	DNP	HRM	IIR	WUE
BIE	**0.816**					
DE	0.795	**0.841**				
DNP	0.789	0.801	**0.834**			
HRM	0.713	0.788	0.765	**0.795**		
IIR	0.804	0.813	0.809	0.781	**0.821**	
WUE	0.721	0.743	0.755	0.712	0.759	**0.782**

Table 4 Testing the hypotheses

Hypothesis		Standardized path coefficient (β)	D. E.	(Bootstrapping)	p-valor
Hip1	HRM =>Business success	0.183	0.062	2948	0.003
Hip2	DNP =>Business success	0.214	0.051	4183	0.000
Hip3	BIE =>Business success	0.113	0.067	1689	0.044
Hip4	WUE =>Business success	0.135	0.045	3013	0.003
Hip5	IIR =>Business success	0.310	0.054	5778	0.000

for business success was 0.760, which is considered acceptable according to Cohen's (1988) suggestion. Finally, the standardized root mean square residual (SRMR) was 0.073.

5 Discussion

The results of this investigation are supported by the findings of a number of other studies that have been done previously. For example, the acceptance of the first hypothesis shows that new product design as a dimension of EEA plays its important role in the growth of microenterprises' business success, and this result is contrary to that shown by Pallathadka et al. [22] who mention that product design is considered as an expense and not as an investment. Furthermore, with respect to the human resource management dimension of ESE and the business success of microenterprises, the study has shown that it affects positively due to the adaptation of human resource management styles prevalent in microenterprises, and this result is similar to what is shown by Zhang et al. [31] concluding that the goals of the plans are properly defined and executed within the timelines in all cases, leading to better business success of microenterprises. Moving on to the next dimension, initiating investor relations has been shown to significantly influence entrepreneurial business success, showing a higher path coefficient among all latent constructs. Due to the uncertainty and dynamic conditions of Peruvian industry, microentrepreneurs have to play the important role of initiating relationships with investors in order to survive in the competition or maneuver the growth of their facilities in a positive way. This role of initiating relationships with investors by entrepreneurs eventually leads them to improve in the level of business success [28]. On the other hand, the acceptance of the fourth hypothesis with a narrow margin shows that although building an innovative environment plays its role in scaling up the business success of microenterprises, microenterprises in Peru put less emphasis on the dimension of building an innovative environment. The reason is that the revenues of micro- and small enterprises are small, and therefore innovation is not considered as part of the budget of the five hypotheses; this relationship turned out to be the weakest. Similarly, [27] propose that innovation has a direct and significant influence on organizational business success. Finally, according to the working under stress dimension of ESE, it has shown that it significantly influences microenterprises' business success. The reason for this result is that nowadays it is necessary for the entrepreneur to have soft skills to improve communication in times of stress [31]. In short, it is concluded that EEA is present in Peruvian micro- and small enterprises and positively affects the business success of microenterprises. Similarly, this study suggests that it is important to develop the five dimensions of EEA in order to achieve a reasonable level of business success.

Although this study makes a number of useful contributions, it does so under certain constraints. The AEE scale, for instance, uses self-reported metrics adapted from a reliable study. However, it may be more instructive to utilize objective measurements, such as revenue growth, to gauge business performance. The paucity of data on most microenterprises was the biggest barrier to getting objective metrics of business success in this study. In the absence of objective metrics, [32] argue that it is possible to acquire subjective measurements. Additionally, the five dimensionality of ESE was derived from a single construct in this study, which did not compromise the reliability or validity of the scale. However, in the future, different constructs that may affect the business success of micro- and small enterprises can be used to assess the effect on the dependent variable. Information from other countries can be gathered as well, giving a more complete picture. Entrepreneurial experience enhances entrepreneurship and plays a significant role in entrepreneurs, especially in the case of EEA [21]. Future research can take into account the experience level of entrepreneurs, as those with more experience tend to demonstrate higher levels of EEA and have greater success in the business world.

6 Conclusions

The findings of this research suggest that new and existing businesses should prioritize spending money on increasing their EEA. Investors should prioritize the EEA of would-be business owners by requiring them to take certain actions. In addition, would-be business owners should aim for a reasonable score across all five dimensions if they want to take their enterprises to the next level of commercial success. Concerning microenterprises, the state can set up some mechanisms to evaluate the owners who score high on the five dimensions of ESE, and micro- and small-business owners may invest in their workers' entrepreneurial startup ability by helping them build an enterprise startup application (ESE), by providing them with educational training courses on product design, human resource development, relationship building, and entrepreneurship.

EEA is present in micro- and small enterprises in Peru and positively affects the performance of MSEs; similarly, this study suggests that it is important to develop the five dimensions of EEA to achieve a reasonable level of performance. As such, the findings of this research may be used as a blueprint for the administration of startups and other developing businesses, so that investment should be made in building and improving the level of ESE.

Investors should prioritize the EEA of those who want to start enterprises by enforcing certain measures. In addition, if would-be business owners want to see significant growth in their enterprises, they should think about how important it is to do well in all five dimensions. When it comes to MSEs, the state can set up some mechanisms to evaluate the business owners who have high scores across all five dimensions of ESE, and those business owners can then work to cultivate the ESE of their future key employees so that they can use them to launch their own ventures. This can be done, for example, through training programs focused on new product design, human resource development, fostering relationships with investors, and so on. The AEE scale relies on self-reported measurements, which is a disadvantage because it was adapted from previously published research; nonetheless, it may be more effective to employ objective measures, rather than self-reported ones, to gauge performance.

In this investigation, the absence of data for most MSEs was the biggest obstacle to establishing objective measurements of performance. Subjective metrics can be produced if objective ones are unavailable. In addition, While the use of a single ESE construct with five dimensions did not compromise the validity of the scale in this study, future research may wish to incorporate multiple constructs that may impact the performance of MSEs, such as technological turbulence, environmental uncertainty, and government support.

Finally, future studies can take into account the fact that entrepreneurs with more experience are more likely to demonstrate superior ESE and, by extension, higher levels of ED. This is because prior experience fosters entrepreneurship and plays a valuable role in entrepreneurs, particularly in the case of ESE. Finally, this study's central finding is that micro- and small-business owners in four Peruvian areas benefit from raising their levels of entrepreneurial self-efficacy.

References

1. Marulanda-Valencia, F. Ã., & Valencia-Arias, J. A. (2019). EvoluciÃ³n y tendencias investigativas en autoeficacia emprendedora: un análisis bibliométrico. *Estudios Gerenciales, 35*(151), 219–229. https://doi.org/10.18046/j.estger.2019.151.3277
2. Thommandru, A., Espinoza-Maguiña, M., Ramirez-Asis, E., Ray, S., Naved, M., & Guzman-Avalos, M. (2021). Role of tourism and hospitality business in economic development. *Materials Today: Proceedings*. https://doi.org/10.1016/j.matpr.2021.07.059.
3. Ramirez-Asis, E., Maguina, M. E., Infantes, S. E., & Naranjo-Toro, M. (2020). Emotional intelligence, competencies and performance of the university professor: Using the SEM-PLS partial least squares technique. *Revista Electrónica Interuniversitaria de Formación del Profesorado, 23*, 99–114. https://doi.org/10.6018/reifop.428261
4. Morales, F. M., Gimenez, J. M., & Morales, A. M. (2018). Relaciones entre autoeficacia emprendedora y otras variables psico-educativas en universitarios. *European Journal of*

Investigation in Health, Psychology and Education, 8(2), 91–102. https://doi.org/10.30552/ejihpe.v8i2.247
5. Fuster-Guillén, D., Ocaña-Fernández, Y., Salazar, D. E., & Ramirez, E. H. (2020). Desarrollo humano e integración familiar: Estudio desde el servicio integral del adulto mayor en Perú. *Revista Venezolana de Gerencia (RVG), 25*(90), 477–490. https://doi.org/10.37960/rvg.v25i90.32392
6. Garcia, M. E., Ramírez, E. H., Lirio, F. A., & Norabuena, R. P. (2015). La formación académica y el espíritu empresarial de los estudiantes de la Escuela de Administración de la Universidad Nacional Santiago Antúnez de Mayolo. *conocimiento para el desarrollo, 6*(1), 89–96. https://revista.usanpedro.edu.pe/index.php/CPD/article/view/108
7. Valencia-Arias, J. A., & Marulanda-Valencia, F. Á. (2019). Evolución y tendencias investigativas en autoeficacia emprendedora: un análisis bibliométrico. *Estudios Gerenciales, 35*(151), 219–232. https://doi.org/10.18046/j.estger.2019.151.3277
8. Manosalvas, L. O. (2017). *Relación de la inteligencia emocional con la intención de emprendimiento y la autoeficacia emprendedora*. Pontificia Universidad Catolica del Peru-CENTRUM Catolica (Perú).
9. Novelo, A. F., Carrillo, A. L. B., & López, S. C. (2016). El impacto de la autoeficacia emprendedora en los resultados de innovación de los emprendedores del sector de software en Yucatán. *Red Internacional de Investigadores en Competitividad, 10*, 559–573. https://www.riico.net/index.php/riico/article/view/1334
10. Yating, Y., Mughal, N., Wen, J., Ngan, T. T., Ramirez-Asis, E., & Maneengam, A. (2022). Economic performance and natural resources commodity prices volatility: Evidence from global data. *Resources Policy, 78*, 102879. https://doi.org/10.1016/j.resourpol.2022.102879
11. Torres-Coronas, T., & Vidal-Blasco, M. A. (2019). La importancia del control conductual percibido como elemento determinante de la intención emprendedora entre los estudiantes universitarios. *Revista Universidad y Empresa, 21*(37), 108–135. https://doi.org/10.12804/revistas.urosario.edu.co/empresa/a.6522
12. Shaikh, A. A., Lakshmi, K. S., Tongkachok, K., Alanya-Beltran, J., Ramirez-Asis, E., & Perez-Falcon, J. (2022). Empirical analysis in analysing the major factors of machine learning in enhancing the e-business through structural equation modelling (SEM) approach. *International Journal of System Assurance Engineering and Management, 13*(Suppl 1), 681–689. https://doi.org/10.1007/s13198-021-01590-1
13. McGee, J. E., & Peterson, M. (2019). The long-term impact of entrepreneurial self-efficacy and entrepreneurial orientation on venture business success. *Journal of Small Business Management, 57*(3), 720–737. https://doi.org/10.1111/jsbm.12324
14. Ríos-Manríquez, M., López-Mateo, C., & Sánchez-Fernández, M. D. (2019). El capital humano como factor clave en el desempeño empresarial: Una visión desde el empowerment. *Revista de Gestão e Secretariado, 10*(3), 69–88. https://doi.org/10.7769/gesec.v10i3.914
15. Leyva, G. (2018). Indicadores de desempeño empresarial para medir la calidad de las estrategias financieras. *Cofin Habana, 12*(1), 58–75. http://www.cofinhab.uh.cu/index.php/RCCF/article/view/278
16. Higuerey, A., & Armas, R. (2021). Evaluación de la supervivencia de las empresas de comunicación ecuatorianas a través de la eficiencia. *Revista Ibérica de Sistemas e Tecnologias de Informação, E40*, 512–525. http://www.risti.xyz/issues/ristie40.pdf
17. Alvarez-Aros, E. L., & Álvarez Herrera, M. (2018). Estrategias y prácticas de la innovación abierta en el rendimiento empresarial: Una revisión y análisis Bibliométrico. *Investigación administrativa, 47*(121), 1–29. https://doi.org/10.35426/IAv47n121.04
18. Wang, C. H., Chang, C. H., & Shen, G. C. (2015). The effect of inbound open innovation on firm business success: Evidence from high-tech industry. *Technological Forecasting and Social Change, 99*, 222–230. https://doi.org/10.1016/j.techfore.2015.07.006
19. Khalil, M., Khan, M., Zubair, S., Saleem, H., & Tahir, S. (2021). Entrepreneurial self-efficacy and small business success in Pakistan. *Management Science Letters, 11*(6), 1715–1724. https://doi.org/10.5267/j.msl.2021.2.011

20. Asandimitra, N., & Kautsar, A. (2017). Financial self-efficacy on women entrepreneurs success. *International Journal of Academic Research in Business and Social Sciences, 7*(11), 293–300. https://doi.org/10.6007/IJARBSS/v7-i11/3459
21. Tian, G., Wang, J., Zhang, Z., & Wen, Y. (2019). Self-efficacy and work business success: The role of work engagement. *Social Behavior and Personality: An International Journal, 47*(12), 1–7. https://doi.org/10.2224/sbp.8528
22. Pallathadka, H., Ramirez-Asis, E. H., Loli-Poma, T. P., Kaliyaperumal, K., Ventayen, R. J. M., & Naved, M. (2021). Applications of artificial intelligence in business management, e-commerce and finance. *Materials Today: Proceedings*. https://doi.org/10.1016/j.matpr.2021.06.419
23. Zisser, M. R., Johnson, S. L., Freeman, M. A., & Staudenmaier, P. J. (2019). The relationship between entrepreneurial intent, gender and personality. *Gender in Management, 34*(8), 665–684. https://doi.org/10.1108/GM-08-2018-0105
24. Bakar, M. S., Ramli, A. B., Ibrahim, N. A., & Muhammad, I. G. (2017). Entrepreneurial self-efficacy dimensions and higher education institution business success. *International Journal of Management Studies (IJMS), 24*(1), 119–137. http://www.e-journal.uum.edu.my/index.php/ijms/article/view/10479
25. Keane, S. F., Cormican, K. T., & Sheahan, J. N. (2018). Comparing how entrepreneurs and managers represent the elements of the business model canvas. *Journal of Business Venturing Insights, 9*, 65–74. https://doi.org/10.1016/j.jbvi.2018.02.004
26. Antoncic, B., Antoncic, J. A., & Aaltonen, H. M. (2016). Marketing self-efficacy and firm creation. *Journal of Small Business and Enterprise Development, 23*(1), 90–104. https://doi.org/10.1108/JSBED-07-2015-0093
27. Herath, H. M., & Mahmood, R. (2014). Dimensions of entrepreneurial self-efficacy and firm business success. *Global Journal of Management and Business Research, 14*(4), 22–30. https://journalofbusiness.org/index.php/GJMBR/article/view/1266.
28. Raza, M., Wisetsri, W., Chansongpol, T., Somtawinpongsai, C., & Ramírez, E. H. (2020). Fostering workplace belongingness among employees. *Polish Journal of Management Studies*, 22(2), 428–442. https://doi.org/10.17512/pjms.2020.22.2.28
29. Supo, J., & Zacarias, H. (2020). *Metodología de la Investigación Científica: Para las Ciencias de la Salud y las Ciencias Sociales*. Sincie.
30. Sarstedt, M., & Cheah, J. H. (2019). Partial least squares structural equation modeling using SmartPLS: A software review. *Journal of Marketing Analytics, 7*(3), 162–202. https://doi.org/10.1057/s41270-019-00058-3
31. Zhang, J., Raza, M., Khalid, R., Parveen, R., & Ramírez-Asís, E. H. (2021). Impact of team knowledge management, problem solving competence, interpersonal conflicts, organizational trust on project business success, a mediating role of psychological capital. *Annals of Operations Research, 1-21*. https://doi.org/10.1007/s10479-021-04334-3
32. Khalid, R., Raza, M., Sawangchai, A., Allauca, W. J., & Huerta, R. M. (2021). Women entrepreneurial innovative behavior: The role of lean start-up and business coaching. *Studies of Applied Economics, 39*(8), 1–19. https://doi.org/10.25115/eea.v39i8.5132

The Connection Between Resilience and Professional Success in Peruvian Superior Court Workers

Mercedes Huerta-Soto, Nelson Cruz-Castillo, Jose Rodriguez-Kong, Lilia Uribe-Pomachagua, and Edwin Hernan Asis

Abstract The study aimed to explore the link between resilience and job performance in the Superior Court of Justice of Ancash, Peru. This research followed a quantitative, correlational approach and employed a non-experimental, cross-sectional design. Data was collected through surveys from 415 male and female employees in various roles within the court. The results showed that 45.1% had a moderate level of resilience, while 46.5% displayed a moderate level of professional success. Similarly, 53.3% demonstrated a moderate level of innovation. Satisfaction levels were reported at 45.8% moderate, efficiency at 36.1% high, growth at 41.7% moderate and productivity quality at 46.5% moderate. The analysis revealed a significant positive relationship between resilience and job performance, supported by a Chi-square value of 136.192 (with 4 degrees of freedom and a significance level of 0.000). It is concluded a strong correlation between these variables in the Superior Court of Justice of Ancash, Peru.job performance variables in the context of the Superior Court of Ancash, Peru.

Keywords High court of justice · Professional success · Resilience · Public servants · Work behaviour

M. Huerta-Soto (✉) · L. Uribe-Pomachagua
Universidad Nacional Santiago Antúnez de Mayolo, Huaraz, Peru
e-mail: mhuertas@unasam.edu.pe; luribep@unasam.edu.pe

N. Cruz-Castillo
Universidad Nacional Mayor de San Marcos, Lima, Peru
e-mail: ncruzca@unmsm.edu.pe

J. Rodriguez-Kong
Universidad Señor de Sipán, Chiclayo, Peru
e-mail: rodriguezka@crece.uss.edu.pe

E. H. Asis
Universidad San Pedro, Chimbote, Peru

J. M. K. P. et al. (eds.), *Business Resilience and Digital Technology in the Post-Pandemic Era*, Contributions to Environmental Sciences & Innovative Business Technology, https://doi.org/10.1007/978-3-031-48075-1_2

1 Introduction

Oliveira and Sala [1] point out that various studies carried out worldwide have shown the importance that there are people with a series of factors or causes who do not have an adequate management of their emotions and attitudes when faced with a problem or in other cases do not know how to handle unpleasant or conflictive situations, generating greater problems of communicational and social interaction between co-workers. Resilience explains why some people respond better to adversity than others and then argues that there are emotional competencies (autonomy, self-efficacy, self-evaluation and capacity for interaction, among others), cognitive competencies (creativity, humour, initiative and problem-solving) and spiritual, ethical and moral capacities necessary for resilient behaviour. In the presidential office of Lima, they studied stress and resilience at work in workers; they found a significant positive relationship between the variables in the staff of the state agency because Spearman = 0.694 ($p = 0.000$) [2].

García [3] argued that due to the work overload in the processing of files until the final resolution is issued in the courts of the Superior Court of Ancash, the organization often has a very low resilience because it does not have a talent for resistance in the face of crisis or uncertain and conflictive situations or learning of things that can be used, as a way to progress and to establish routes to success as a way to survive, through resilience, not limiting the crisis in the judicial entity only to overcoming the temporary attention to the files but to personal and institutional progress. There are activities that generate work stress such as the overload of court files, becoming adverse events, events that produce work stress, when civil servants and workers do not practice resilient behaviour. The important thing is to adapt and develop positively in the face of adverse circumstances.

In public servants in Puno, work stress, generated by the lack of resilient behaviour, became a very serious problem in the institutional environment, affecting the place where the people of the Superior Court of Puno live and work as they know the demands within the work, each time they demand more and more demanding changes, producing personal and institutional effects, such as a negative impact on the fulfilment of processes, activities and tasks [4]. The lack of resilient behaviour is a widespread problem in the superior courts that generates a professional success with a lack of compliance with the work standards established for the judiciary in the different courts that comprise a superior court. These circumstances have a set of factors that condition this situation produced by a series of deficiencies in work behaviour and the inconveniences that arise in the operators of justice. This situation would be repeating itself in the professional success of the workers of the Superior Court of Ancash, as a result of the interaction between managers and workers who fulfil the functions of justice operators, who neglect the effects of the intervention that resilience should emerge, valuing the positive role they play in the interrelation between co-workers, with the heads of areas and with the public user by putting into practice the positive psychic behaviour of the human being, i.e. they would not be feeling as their own their feelings, moods and attitudes related to their work activity in order to achieve a competitive advantage.

Sibalde et al. [5] argue that, in order to define resilience, much importance is given to temperament, as an element for the relationship between people, as biological substances and psychological factor constitute the personality. This situation would be made worse when the justice operator does not work efficiently, without greater motivation and commitment to the court that comprises the organization of the superior court; demonstrating low motivation and a deficient level of commitment to the institution of the State, in this order of ideas, the results would present anomalies due to poor professional success; furthermore, he would demonstrate little competitiveness or efficiency in relation to other co-workers, without allowing him personal and professional development, presenting anomalies in the fulfilment of his functions, established by the current regulations and the manual of organization and functions of the job position that corresponds to him [6]. Under these conditions, there would be obstacles in the communicative interaction and coordination of processes, activities of tasks of the fulfilment of the judicial processes, established for each legal act in the Supreme Court, when there is a lack of resilient behaviour in all the work order.

This is reflected in the dissatisfaction of the user population, in a negative way, because their needs and expectations of justice in the solution of their problems are not met. Because most workers are often unable to detect the causes of problems because they do not use resilience techniques [7]. In this situation, they are rarely able to manage their emotions and their thoughts, and they use them without control, despite having a negative charge, taking into account that a negative thought triggers negative emotions without facing adversity positively. In other cases, they do not know how to maintain passivity at work with a lot of pressure, they cannot cope with difficult situations in their daily life and they do not keep calm in adverse moments and are not able to be calm and focused in the face of confusion and chaos. They do not demonstrate optimistic realism and positive thinking, hoping for the best that characterizes resilient people, learning lessons from the behaviour of others to grow within adversity and finding opportunities for positive learning and individual development for their benefit.

Sometimes, workers who are more able to cope with problems do not know how to read and understand the emotions of others and confront the negative consequences; there are workers who are unhappy and do not have the ability to show empathy with other co-workers, who present cases of depression and who show their own needs in a defensive way; on the other hand, the resilient person is more empathetic by taking life in a different way and facing life in a positive and optimistic way, making them a happier person. There are people with different characteristics because they are not all the same: because one person is motivated by a challenge and another person by opportunities for change; and because they are not resilient, they do not know how to look for new ways to achieve conformity in their life, without keeping their motivation at a high level. In addition, they collapse in the face of problems, because they let themselves be carried away by a series of reproachful thoughts and are always questioning the negative situations that always happen to them, without applying strategies of how to get out of this conflictive situation. In other words, resilience is latent in every person, as is the lack of practice of resilient behaviour. For the reasons described and explained above, it has been decided to propose research questions, formulated at a general level. How is resilience related to professional success in the High Court of Justice of Ancash?

2 Review of the Literature

Resilience for [8] argue that we live the act of exchanging messages through conversation in the cultural context in which we develop our personal development. As proposed by Chiavenato and González [9] from the cognitive point of view, humans structure "design environments", where human reasoning can overcome the biological brain, by interacting with the complex universe, both physically and socially. Inter-relationships with the world can set limits, while at the same time, they can enhance problem-solving abilities and help with stress management and coping skills when critical moments arise. In this respect, [10] held in the 1930s the position that the use of language generates profoundly the development of cognition, maintaining that discourse produces a relationship between discourse and learning and social experience and proposing the concept "Zone of proximal development", to name the need that the child has on the orientation that the other person provides and to triumph over the activities, which the development in life imposes.

From the perspective of resilience, it is called the adult person who motivates the child or adolescent to achieve a level of where to reflect on the situation of the moment, as well as to have, as well as to have more control of their acts and their conscience. García-Contreras et al. [11] concluded that in order to acquire the mechanism of language, there is no possibility of achieving it if there is no systemic support provided by the social world in combination with the ability to acquire learning. He reaffirms the need for the presence of another person, for the acquisition of language, equivalent to expressing that the human being needs another person, to enter the cultural context [12].

Professional success [9] is the "efficiency shown by the staff working in the organization, very necessary for the company, acting the individual with a good job and job satisfaction". In this sense it can be said that job performance consists of the combination of its activity and its effects, so it will be necessary to make a calculation of attention to action. Thus, performance demonstrates the worker's capabilities in generating successful performance in the job, in the shortest time with the least effort, with productive effectiveness.

Bocco [13] in his theory maintains that "competences are skills or objective behaviours provided by the person in the job in compliance with their assigned functions in an efficient, effective and satisfactory manner". Further on, research establishes three types of explicit competences: "Physical, occupational and generic; to be used in accordance with institutional objectives". On the other, Mejía and Benítez [14] stated that human resources should be evaluated in their performance, using various techniques, instruments and strategies, criteria and methods, to assess the worker's performance (p. 55). The methods established by the aforementioned authors were:

Human resource planning:

According to the available data of an entity or a company, the general HR needs are calculated and what corresponds to the worker and specifically to the managers.

Selection and recruitment: It will be necessary to systematize the evaluation of performance, in order to foresee the performance of the applicants for each job position.

Any evaluation of (HR) must comply with the actions of training and professional and personal development of the entrant, taking advantage of their capabilities for success and minimizing their shortcomings [15], posed as "a set of feelings, favourable or unfavourable, that allows the employee to perceive their work, which is expressed in various work attitudes" (p. 203). The theory relates to the essence of the job and its environment, such as organization, work teams, supervision and monitoring. In this sense, the authors state that job satisfaction is a perception of pleasure or pain, very divergent from the purposes and goals and thoughts of human behaviour; these attitudes allow managers to make a projection into the future, the future outcome of work activities.

3 Methodology

The research was applied or applicative because its objective was to investigate and consolidate knowledge in order to apply it in practice, enriching scientific and cultural progress. It was correlational because it studied the relationship of significance between variables and the dimensions of the study [16]. The population consisted of 415 male and female workers, appointed and contracted in the Superior Court of Justice of Ancash. With the following breakdown: 43 judges from the 276 system; 138 administrative and judicial workers from the 728 system; 49 judicial assistants, hired for 1, 3 and 6 months of service. The structured questionnaire was used, which is a means or sheet of questions specially prepared, based on the objectives, the technique and the table of operationalization of variables, taking into account the indicators converted into structured questions with polytomous alternatives [17]. In the case of applying a validated instrument, the application and data processing processes were taken into account in the methodology and the treatment of the results.

4 Results

In Table 1 and Fig. 1, we can see that in the perception of the workers of the Superior Court of Justice of Ancash, Peru, on resilience, 31.3% mentioned a high level of resilience, 45.1% a medium level and 23.6% a low level of resilience.

Table 1 Symmetric measurements

		Value	T Approximate[a]	Approximate Significance
Interval by interval	Pearson's R	0.794	15,538	0.000[b]
Ordinal by ordinal	Spearman's correlation	0.790	15,352	0.000[b]
N of valid cases		415		

[a]Use of the asymptotic standard error that assumes the null hypothesis
[b]Based on normal approximation

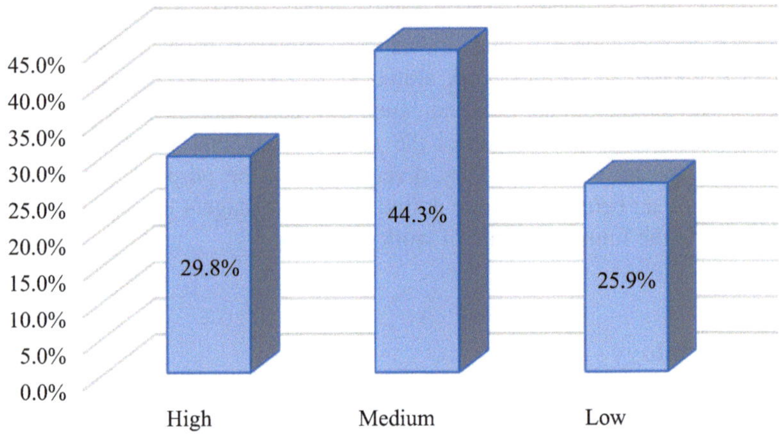

Fig. 1 Levels of resilience

Figure 2 show that 33.3% of the workers of the Superior Court of Justice of Ancash, Peru, had a high level of job performance, 46.5% had a medium level of job performance and 20.1% had a low level of job performance.

$$\text{Chi Cuadrado} = 136{,}192 \qquad \text{gl.} = 4 \qquad \text{sig.} = 0.000$$

There is sufficient evidence to conclude that there is a significant positive relationship between the variable's resilience and job performance in the Superior Court of Justice of Ancash, Peru (Chi-square, $^{***}p = 0.000$). Furthermore, according to Spearman's correlation test = 0.790 ($^{***}p = 0.000$), it is concluded that there is a direct, strong and significant correlation between the variables.

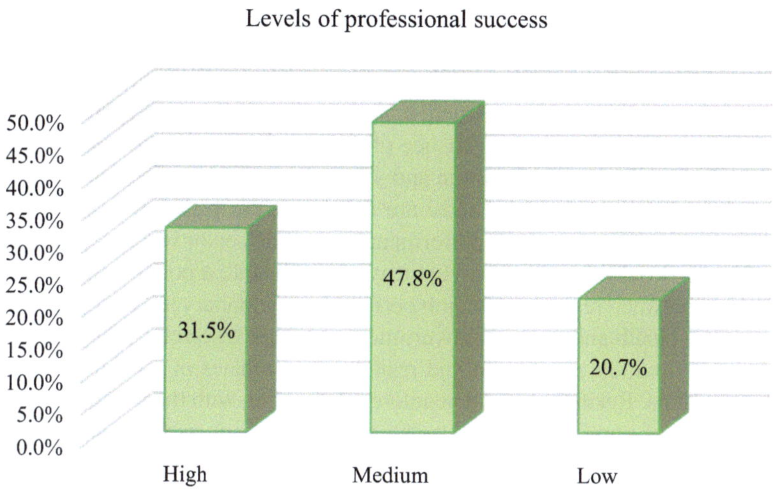

Fig. 2 Job performance

5 Discussion

The relationship between resilience and professional success in the Ancash High Court of Justice was determined. From the statistical results obtained, it can be affirmed that the workers of the Superior Court of Ancash, Peru, perceive resilience, to be 45.1% of medium level and 46.5% of medium-level professional success. It is related to the paper of Talavera et al. [18] who studied the evaluation of burnout and resilience-engagement factors in teachers; the low voluntary participation reflects the lack of interest, to demonstrate factors that are related to burnout in the group, although the effects cannot be generalized. It concludes that resilience decreases vulnerability to burnout. Theoretically, resilience is "the generalized disposition that causes a person, group or community to prevent, reduce or overcome the damaging effects of a disaster" [19]. Therefore, there is a direct relationship between the results found with the outputs found by Koontz et al. [10], who argue that vulnerability is reduced by resilience.

The statistical results obtained indicate that the workers of the Ancash Superior Court of Justice perceive innovation to be high for 22.9% and medium for 53.3%. The results obtained from this research are related to the paper by Alzugaray et al. [20] who investigated about resilience and job performance in the administrative workers of the Autonomous University of Ica, with the objective of explaining the relationship between resilience and job performance in the administrative workers of the Autonomous University of Ica. It concluded that the level of resilience is related to job performance in workers in the administrative of the University Centre for Studies. Theoretically, "Innovation is about breaking routines through increased participation and commitment" [21]. As can be verified, there is a direct relationship between the specific objective achieved in the present research with the paper of

Ayllón [22], who concluded that there was a direct relationship between the level of resilience and the job performance of the administrative workers of the Universidad Autónoma de Ica.

From the statistical results obtained, it can be affirmed that in the perception of the workers of the Superior Court of Justice of Ancash, Peru, surveyed on satisfaction, 31.9% had a high level of satisfaction and 45.8% a medium level of satisfaction. The results obtained in the present study are related to the paper of Loan [23], who researched on resilience and job performance of public sector workers in Huancayo, concluding that public workers in Huancayo demonstrate a positive and significant relationship between resilience and job performance: judiciary, education, provincial municipality, health and regional government. Another research by Ruvalcaba et al. [24] focused on adult attachment and resilience in inmates in a penitentiary establishment in Lima. Resilience had a negative relationship with the number of children and the length of imprisonment but positive with the frequency of visits, as well as being higher in those who had access to higher education and belonged to a medium-security ward. From a theoretical point of view, job satisfaction is a positive attitude that a person feels in relation to their work according to their own feelings, state of mind and attitudes. The direct relationship between the specific objective achieved in this research can be demonstrated with the paper of Loan [23], who concluded that workers in Huancayo show a positive and significant relationship between resilience and job performance in the public sector.

From the statistical results obtained, it can be assured that the perception of the workers of the Superior Court of Justice of Ancash-Peru, about efficiency, for 36.1% is high efficiency level and 37.5% medium efficiency level. The results found in the present research are related to the paper of Shaikh et al. [25], who developed a paper on work stress and work resilience in the workers of the presidential office, concluding that there is a significant relationship between job stress and job resilience in workers, with a Spearman's Rho $= 0.644^*$ and $p = 0.000$, having a moderate and positive correlation between the variables. Also, [26]) developed a paper on organizational commitment and job performance of the staff of the logistics area, Superior Court of Justice, and concluded that there is a relationship between organizational commitment and job performance. As a theory, efficiency according to [27]) is a criterion used to know how to operate with fewer resources in order to achieve a single or the same objective. As can be verified, there is a direct relationship between the specific objective achieved in this research with the paper of Gómez-Baños et al. [28], who reached a conclusion that there is a significant relationship between work stress and resilience in the workers of the presidential office.

From the statistical results obtained, it can be affirmed that 38.2% of the workers of the Superior Court of Justice of Ancash, Peru, have a high level of growth and 41.7% have a medium level of growth. The results found in this study are related to those of Chen et al. [29], whose paper focused on work stress and resilience in firefighters in the Department of Lambayeque. The results determined that there is a significant relationship between work stress and resilience, indicating that the demands and work pressures that these workers go through generate tolerance to stress and strengthen the individual's ability to cope with adversity, feeling strengthened by the demands of the job. Furthermore, [30]) affirmed that resilience can be produced at any stage of life, becoming the instrument for overcoming the vulnerable state, in the face of mishaps. Resilience minimizes the negative stressful effect. According to [31]), the theory of growth states that everyone wants to be and have more, never the opposite. From this point of view, they perceive to be oriented towards new realities, which give them the possibility to grow and improve their expectations. It is proven that there is a direct relationship between the specific objective achieved in this research with the paper of Mendieta-Ortega et al. [21], who concluded that the population achieved a low level of work stress and a high level of resilience.

The perception of the surveyed workers on productivity was verified; for 32.6%, it was high level of productivity and 46.5% medium level of productivity. The results found in the research are related to the paper of Thommandru et al. [32], who studied on the procedural burden and its influence on the work performance of the staff of the Third Magistrates' Court of the Superior Court of Justice of La Libertad, coming to the conclusion that there is an average negative correlation between procedural burden and work characteristics. In theory, productivity is a measure of how well an organization's procedures and operations are working. In conclusion, there is a direct relationship between the specific objective achieved in the present research with the paper of Khoshnaw and Alavi [33], who concluded that there is a negative average correlation between procedural burden and work characteristics.

From the statistical results obtained, it can be affirmed that 31.9% of the workers of the Superior Court of Ancash, Peru, perceived quality as high, 47.2% as medium and 20.8% as low. The results obtained are related to the study of Marulanda-Valencia and Valencia-Arias [34], who focused on the influence of managerial leadership on the professional success of workers in the administration and finance unit of the Superior Court of Justice of Santa. They concluded that managerial leadership has a significant ($p = 0.002 < 0.01$) direct and moderate ($r = 0.515$) influence on the job performance of workers. Theoretically, Koontz et al. [10] affirmed that quality in professional success is the growth achieved in activities with a minimum of error, consisting of the production of a product or the provision of a service in accordance with the standards required by the client. It is proven that there is a direct relationship between the specific objective achieved in the research, with the paper of Krijgsheld et al. [35], who mention that resilience is not only important for the individual himself but for society in general, since enriching the resilient factors that help people to face adversity is a key factor in the development of the quality of professional success.

6 Conclusions

The relationship between resilience and professional success in the Superior Court of Justice of Ancash has been determined, with a Chi Square value = 136.192 and a degree of freedom gl. = 4 sig. = 0.000; there is sufficient evidence to conclude that there is a significant positive relationship between the variable's resilience and professional success.

The relationship that exists between resilience and innovation in professional success in the Superior Court of Justice of Ancash is described, with the value of Chi square = 94.050 and a degree of freedom gl. = 4, sig. = 0.000, and it can be verified that there is a positive significant relationship between the variables resilience and job performance innovation. According to Spearman's correlation = 0.654 (***p = 0.000), there is a direct, strong and significant correlation between the variables.

The relationship between resilience and job performance satisfaction in the Superior Court of Justice of Ancash is indicated, with the Chi square value = 112,572 and a degree of freedom gl. = 4, sig. = 0.000; it can be verified that there is a significant positive relationship between the variable's resilience and job performance satisfaction. Also, according to Spearman's correlation test = 0.732 (***p = 0.000), it is concluded that there is a direct, strong and significant correlation between the variables.

The relationship that exists between resilience and the efficiency of professional success in the Superior Court of Justice of Ancash has been determined, with the Chi-square value = 87.454 and a degree of freedom gl. = 4, sig. = 0.000; it can be verified that there is a positive significant relationship between the variable's resilience and the efficiency of professional success. Likewise, according to Spearman's correlation = 0.691 (***p = 0.000), it is concluded that there is a direct, strong and significant correlation between the variables.

The relationship that exists between resilience and the growth of job performance in the Superior Court of Justice of Ancash is specified, with a Chi-square value = 98.530 and gl. = 4, sig. = 0.000; it can be shown that there is sufficient evidence to conclude that there is a positive significant relationship between the variable's resilience and the growth of job performance in the Superior Court of Ancash, Peru. Furthermore, according to Spearman's test = 0.691 (***p = 0.000), it is concluded that there is a direct, strong and significant correlation between the variables.

The relationship that exists between resilience and productivity of professional success in the Superior Court of Justice of Ancash has been identified, with the Chi-square value = 123,491 and a degree of freedom gl. = 4, sig. = 0,000; it is possible to contrast the existence of sufficient evidence to resolve that there is a positive significant relationship between the variable's resilience and productivity of professional success. And according to Spearman's correlation test = 0.749 (***p = 0.000), it is decided that there is direct, strong and significant correlation between the variables.

The relationship that exists between resilience and the quality of professional success in the Superior Court of Justice of Ancash is clear; according to the obtained value of Chi-square = 126,324 and a gl. = 4, sig. = 0.000, it can be verified that there is a significant positive relationship between the variable's resilience and the quality of professional success in the Superior Court of Ancash, Peru. Also, according to Spearman's correlation test = 0.746 (***p = 0.000), it can be concluded that there is a direct, strong and significant correlation between the variables.

References

1. Oliveira, L. G., & Sala, R. (2021). Importancia de la resiliencia individual y organizacional para la salud laboral del personal sanitario. *Revista de la Asociación Española de Especialistas en Medicina del Trabajo, 30*(2), 151–160. https://scielo.isciii.es/scielo.php?script=sci_arttext&pid=S1132-62552021000200151
2. Yating, Y., Mughal, N., Wen, J., Ngan, T. T., Ramirez-Asis, E., & Maneengam, A. (2022). Economic performance and natural resources commodity prices volatility: Evidence from global data. *Resources Policy, 78*, 102879. https://doi.org/10.1016/j.resourpol.2022.102879
3. García, S. (2022). *La resiliencia de las organizaciones: El secreto está en el sistema*. Libros de Cabecera.
4. Sharma, A., Agrawal, R., & Khandelwal, U. (2019). Developing ethical leadership for business organizations: A conceptual model of its antecedents and consequences. *Leadership & Organization Development Journal., 40*(6), 712–734. https://doi.org/10.1108/LODJ-10-2018-0367
5. Sibalde, I. C., Sibalde, M. D. A., Da Silva, A. D., Scorsolini-Comin, F., Brandão, W., & Meirelles, E. M. L. (2020). Factores relacionados con la resiliencia de adolescentes en contextos de vulnerabilidad social: revisión integradora. *Enfermería Global, 19*(59), 582–625. https://doi.org/10.6018/eglobal.411311
6. Morgan, J. (2020). La resiliencia: habilidad esencial para hacerle frente a la cuarta revolución industrial. *Revista nacional de Administración, 11*(1), 21–31. https://doi.org/10.22458/rna.v11i1.2970
7. Ramirez-Asis, E., Maguina, M. E., Infantes, S. E., & Naranjo-Toro, M. (2020). Emotional intelligence, competencies and performance of the university professor: Using the SEM-PLS partial least squares technique. *Revista Electronica Interuniversitaria de Formación del Profesorado, 23*, 99–114. https://doi.org/10.6018/reifop.428261
8. Gamboa-Moreno, L. N., Becerra-Rodríguez, K. G., Lopez-Vergara, Y. I., & Goicochea-Ríos, E. (2021). Nivel de resiliencia del personal de salud frente a la pandemia por Covid-19. *Revista del Cuerpo Médico Hospital Nacional Almanzor Aguinaga Asenjo, 14*, 49–54. https://doi.org/10.35434/rcmhnaaa.2021.14sup1.1170
9. Chiavenato, I., & González, E. A. (2017). *Comportamiento organizacional: la dinámica del éxito en las organizaciones*. McGraw-Hill.
10. Koontz, H., Weihrich, H., & Cannice, M. V. (2020). *Essentials of management-an international, innovation and leadership perspective* (11th ed.). McGraw-Hill Education.
11. García-Contreras, R., Valle-Cruz, D., & Canales-García, R. A. (2021). Selección organizacional: resiliencia y desempeño de las pymes en la era de la COVID-19. *Estudios Gerenciales, 37*(158), 73–84. https://doi.org/10.18046/j.estger.2021.158.4291
12. Rogel-Gutiérrez, E. M., & Urquizo-Villafuerte, J. I. (2019). Aproximación teórica a la resiliencia en las organizaciones financieras. *Revista De Ciencias Sociales, 25*(2), 112–119. https://doi.org/10.31876/rcs.v25i2.27340

13. Bocco, G. (2019). Vulnerabilidad, adaptación y resiliencia sociales frente al riesgo ambiental. Teorías subyacentes. *Investigaciones Geográficas, 100*, e60024. https://doi.org/10.14350/rig.60024
14. Mejía, G. M., & Benítez, F. E. B. (2020). Educación integral para el éxito profesional del Administrador. *Revista GEON (Gestión, Organizaciones y Negocios), 7*(1), 83–95. https://doi.org/10.22579/23463910.176
15. Pico-Saltos, R., Garzás, J., Redchuk, A., & Jaramillo, I. F. (2023). Medición del éxito profesional a graduados de la Universidad Técnica Estatal de Quevedo-Ecuador. *Revista De Ciencias Sociales, 29*(2), 432–443. https://doi.org/10.31876/rcs.v29i2.39986
16. Ñaupas, H., Valdivia, M., Palacios, J., & Romero, H. (2018). *Metodología de la investigación Cuantitativa-Cualitativa y redacción de la tesis* (5th ed.). Ediciones de la U.
17. Hernández, R., & Mendoza, C. (2018). *Metodología de la investigación: Las rutas cuantitativa, cualitativa y mixta*. Mc Graw Hill.
18. Talavera, B., Luceño, L., García, Y., & Martín, J. (2021). Perception of health, resilience, and engagement in Spanish police officers during the COVID-19 pandemic. *Psicothema*. https://doi.org/10.7334/psicothema2021.153
19. Li, H., Jin, X., Liu, J., Feng, D., Xu, W., & Zhou, Y. (2022). Analytical framework for integrating resources, morphology, and function of rural system resilience—An empirical study of 386 villages. *Journal of Cleaner Production, 365*, 132738. https://doi.org/10.1016/j.jclepro.2022.132738
20. Alzugaray, C., Fuentes, A., & Basabe, N. (2021). Resiliencia Comunitaria: una aproximación cualitativa a las concepciones de expertos comunitarios. *Rumbos TS, 16*(25), 181–203. https://doi.org/10.51188/rrts.num25.496
21. Mendieta-Ortega, M. P., Erazo-Álvarez, J. C., & Narváez-Zurita, C. I. (2020). Gestión por competencias: herramienta clave para el rendimiento laboral del talento humano del sector hospitalario. *Revista Arbitrada Interdisciplinaria Koinonía, 5*(10), 287–312. https://doi.org/10.35381/r.k.v5i10.696
22. Ayllón, M. L. (2016). *Resiliencia y Desempeño Laboral en los Trabajadores Administrativos de la Universidad Autónoma de Ica*. Tesis de pregrado, Universidad Autónoma de Ica. http://repositorio.autonomadeica.edu.pe/handle/autonomadeica/564
23. Loan, L. J. (2020). The influence of organizational commitment on employees' job performance: The mediating role of job satisfaction. *Management Science Letters, 10*(14), 3307–3312. https://doi.org/10.5267/j.msl.2020.6.007
24. Ruvalcaba, N. A., Gallegos, J., & Bravo, H. R. (2019). Validez predictiva de las competencias socioemocionales sobre la resiliencia en adolescentes mexicanos. *Diversitas: Perspectivas en Psicología, 15*(1), 89–101. https://doi.org/10.15332/s1794-9998.2019.0015.07
25. Shaikh, A. A., Lakshmi, K. S., Tongkachok, K., Alanya-Beltran, J., Ramirez-Asis, E., & Perez-Falcon, J. (2022). Empirical analysis in analysing the major factors of machine learning in enhancing the e-business through structural equation modelling (SEM) approach. *International Journal of System Assurance Engineering and Management, 13*(Suppl 1), 681–689. https://doi.org/10.1007/s13198-021-01590-1
26. Darvishmotevali, M., & Ali, F. (2020). Job insecurity, subjective Well-being and job performance: The moderating role of psychological capital. *International Journal of Hospitality Management, 87*, 102462. https://doi.org/10.1016/j.ijhm.2020.102462
27. Matabanchoy-Tulcán, S. M., Álvarez-Pabón, K. M., & Riobamba-Jiménez, O. D. (2019). Efectos de la evaluación de desempeño en la calidad de vida laboral del trabajador: Revisión del tema entre 2008-2018. *Universidad y salud, 21*(2), 176–187. https://doi.org/10.22267/rus.192102.152
28. Gómez-Baños, R., Saldaña-Barrientos, S., Orozco-Arellano, M. A., & Rivas-Vega, B. A. (2022). Correlación entre estrés laboral y resiliencia en los médicos residentes de medicina familiar. *Revista mexicana de medicina familiar, 9*(3), 78–85. https://doi.org/10.24875/rmf.21000149

29. Chen, R., Xie, Y., & Liu, Y. (2021). Defining, conceptualizing, and measuring organizational resilience: A multiple case study. *Sustainability, 13*(5), 2517. https://doi.org/10.3390/su13052517
30. Pallathadka, H., Ramirez-Asis, E. H., Loli-Poma, T. P., Kaliyaperumal, K., Ventayen, R. J. M., & Naved, M. (2021). Applications of artificial intelligence in business management, e-commerce and finance. *Materials Today: Proceedings*. https://doi.org/10.1016/j.matpr.2021.06.419
31. Sanin, J., & Salanova, M. (2014). *Satisfacción Laboral: El Camino entre el Crecimiento Psicológico y el Desempeño Laboral en Empresas Colombianas Industriales y de Servicios*. Universitas Psychologica.
32. Thommandru, A., Espinoza-Maguiña, M., Ramirez-Asis, E., Ray, S., Naved, M., & Guzman-Avalos, M. (2021). Role of tourism and hospitality business in economic development. *Materials Today: Proceedings*. https://doi.org/10.1016/j.matpr.2021.07.059
33. Khoshnaw, S., & Alavi, H. (2020). Examining the interrelation between job autonomy and job performance: A critical literature review. *Multidisciplinary Aspects of Production Engineering, 3*(1), 606–616. https://doi.org/10.2478/mape-2020-0051
34. Marulanda-Valencia, F. Ã., & Valencia-Arias, J. A. (2019). EvoluciÃ³n y tendencias investigativas en autoeficacia emprendedora: un análisis bibliométrico. *Estudios Gerenciales, 35*(151), 219–229. https://doi.org/10.18046/j.estger.2019.151.3277
35. Krijgsheld, M., Tummers, L. G., & Scheepers, F. E. (2022). Job performance in healthcare: A systematic review. *BMC Health Services Research, 22*(1), 149. https://doi.org/10.1186/s12913-021-07357-5

Study on the Potential of an Artificial Intelligence-Based Algorithm for Hand Gesture-Controlled Lighting Systems

Lanh-Le Thanh, Hien Thanh Le, Ngoc Son-An Nguyen, Anh-Luong Quoc, and Ky-Hoang Quoc

Abstract Gesture recognition has gained significant traction in the fields of computer vision and machine learning, offering users a natural and intuitive means of interacting with robots and system interfaces without the need for additional equipment. The primary objective of gesture recognition within human-computer interaction (HCI) is to develop systems capable of identifying specific human gestures and utilizing them for conveying information or controlling devices. To accomplish this, hand gesture interfaces based on visual input necessitate swift and highly precise hand detection alongside real-time gesture recognition. This article presents a comprehensive solution utilizing machine learning algorithms that can be universally applied to a diverse range of real-time gesture recognition systems for human-computer interfaces. The proposed system facilitates controlling the LED light, enabling users to switch it on or off based on the recognition of hand gestures.

Keywords Hand gesture recognition · Artificial intelligence · Hand gesture-controlled

1 Introduction

In recent years, advancements in artificial intelligence (AI) and computer vision have enabled the development of hand gesture-controlled devices. One such device is a hand gesture-controlled lighting system that uses AI algorithms to interpret hand movements and adjust the lighting accordingly. This technology offers a hands-free and intuitive way to control lighting, making it a potentially useful solution for

L.-L. Thanh · N. S.-A. Nguyen · A.-L. Quoc · K.-H. Quoc
Faculty of Information Technology, Dong Nai Technology University, Bien Hoa City, Vietnam
e-mail: lethanhlanh@dntu.edu.vn

H. Thanh Le (✉)
Faculty of Electrical and Electronics Engineering, HCMC University of Industry and Trade, Ho Chi Minh City, Vietnam
e-mail: hienlt@hufi.edu.vn

J. M. K. P. et al. (eds.), *Business Resilience and Digital Technology in the Post-Pandemic Era*, Contributions to Environmental Sciences & Innovative Business Technology, https://doi.org/10.1007/978-3-031-48075-1_3

people with physical disabilities or in environments where traditional switches may not be practical [1, 2].

This study aims to investigate the development of a hand gesture-controlled lighting system using an artificial intelligence algorithm. The system will be designed to interpret hand gestures, such as a wave, a fist, or a pinch, and adjust the lighting in response to these gestures. The study will involve the use of a camera and computer vision techniques to capture and interpret hand movements and an AI algorithm to control the lighting system [3, 4]. The paper will focus on exploring the feasibility and effectiveness of this technology, as well as identifying any potential limitations or challenges. Additionally, the study will examine the usability of the system, including its ease of use, user satisfaction, and potential for commercial applications [3, 5, 6]. Overall, this study on a hand gesture-controlled lighting system using artificial intelligence algorithm has the potential to contribute to the development of new and innovative technologies for hands-free and intuitive control of lighting systems [6, 7].

2 Method Research

The design of the gesture recognition system outlined in Fig. 1 involves several key components and steps.

Data acquisition and pre-processing: The system begins by capturing video images of the user's hand using a suitable sensor or camera. This raw data is then pre-processed to enhance its quality and remove any noise or unwanted artifacts. Pre-processing techniques may include image filtering, background subtraction, or image stabilization.

Hand detection and tracking: The next step involves detecting and tracking the user's hand within the video images. Hand detection algorithms analyze the pre-processed data to identify the region of interest corresponding to the hand. Hand tracking techniques then ensure the continuous tracking of the hand's position and movement over time [8-9].

Hand segmentation and feature extraction: Once the hand is detected and tracked, it is necessary to segment it from the background to isolate the hand region for further analysis. Segmentation techniques can employ thresholding, contour extraction, or skin color modeling to separate the hand from the rest of the image. Following segmentation, relevant hand features are extracted to capture the distinctive characteristics of the hand posture or motion. These features may include finger positions, palm orientation, hand shape, or motion trajectories.

Gesture classification: The extracted hand features are then used for gesture recognition. In the case of static gestures (postures held for a short period), the extracted features are typically normalized to remove variations in hand size, orientation, or lighting conditions. The normalized features form an instance vector that is input to a classification algorithm, which assigns the gesture to a specific class or category.

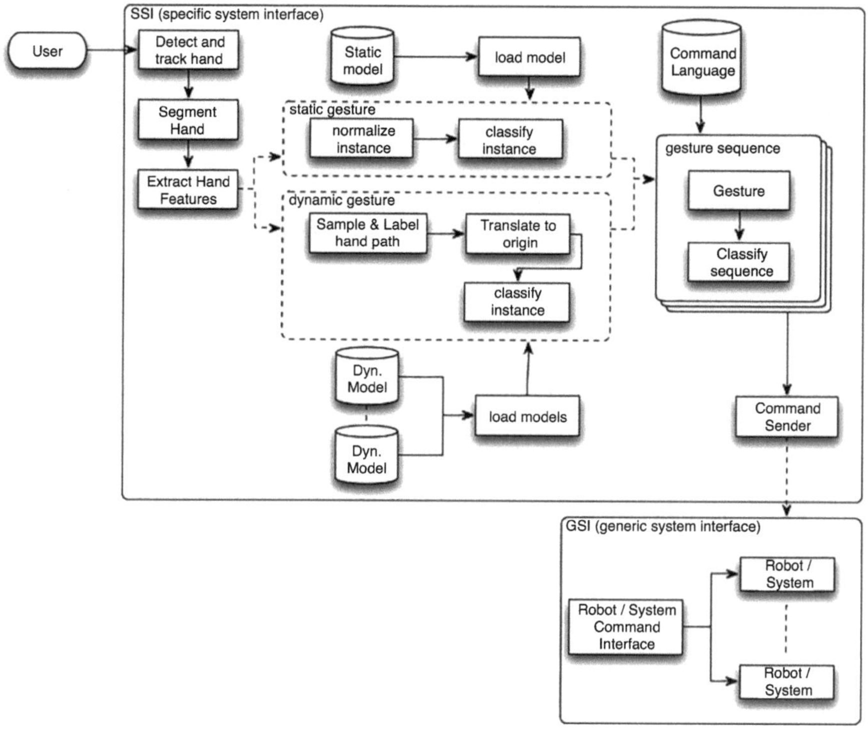

Fig. 1 Vision-based hand gesture recognition system architecture

Dynamic gesture labeling: For dynamic gestures (gestures involving motion), the hand path is labeled using a predetermined alphabet. This labeling process assigns discrete labels to different points along the hand trajectory. The labeled vector of these points is then translated to a common origin to remove the influence of starting positions or orientations.

Sequence recognition: Each detected gesture is processed by a module that accumulates received gestures until a predefined sequence, defined within a command language, is identified. The system waits for a specific sequence of gestures to occur in a particular order. This sequence recognition ensures that the user's intended command is recognized accurately.

Command classification and transmission: Once the sequence is recognized, it is classified into one of the predefined commands from a set number of options. The recognized command represents the user's intended action or instruction. These commands can be transmitted to a Generic System Interface (GSI) to control various robots or systems. The GSI serves as the interface between the gesture recognition system and the target system, enabling the execution of the desired commands. This gesture recognition system enables human-computer interaction by detecting and interpreting hand gestures, allowing users to control systems or robots through intuitive and natural hand movements, shown in Fig. 1.

Hand posture classification heavily relies on hand segmentation and feature extraction within vision-based hand gesture recognition systems. The pre-processing stage is responsible for preparing the input image and extracting features that are later utilized in conjunction with classification algorithms. In the proposed system, the feature vectors are composed of centroid distance values for hand posture classification. The centroid distance signature, a type of shape signature, is represented by the distances between the hand contour boundary points and the hand centroid (xc, yc). This can be calculated using Formula 1:

$$d(i) = \sqrt{(x_i - x_c)^2 + (y_i - y_c)^2}, i = 0, \ldots, N - 1 \quad (1)$$

To obtain a one-dimensional function representing the hand shape, the proposed system employs a feature vector comprising centroid distance values for hand posture classification. This necessitates a critical step in vision-based hand gesture recognition systems, namely, the hand segmentation and feature extraction process. During the pre-processing stage, the input image is prepared, and features are extracted for subsequent use with classification algorithms, as depicted in Formula 2:

$$Z = \frac{(a_{ij} - \overline{a})}{\sigma} \quad (2)$$

In the process of hand posture classification within vision-based hand gesture recognition systems, hand segmentation and feature extraction are crucial steps. The pre-processing stage involves preparing the input image and extracting features for later use with classification algorithms. In this system, the proposed approach for constructing the feature vector entails utilizing centroid distance values for hand posture classification.

To ensure scale invariance, the obtained feature vectors undergo z-normalization. This involves subtracting the mean of the feature vectors and dividing by their standard deviation. By applying this normalization technique, all feature vectors have a mean of zero and a standard deviation of one. The resulting feature vectors are then utilized to train a multi-class support vector machine (SVM). SVM is a supervised machine learning technique widely used in pattern recognition tasks, particularly with high-dimensional data. It selects a small number of boundary feature vectors, known as support vectors, from each class and constructs a linear discriminant function that maximizes the separation between them. The resulting maximum-margin hyperplane is relatively stable, only being modified when support vectors are added or removed during training.

The system needs to be trained or programmed to recognize a set of distinct hand postures or gestures that correspond to specific commands or actions. Here are a few examples of commonly recognized hand postures in gesture recognition systems, shown in Fig. 2:

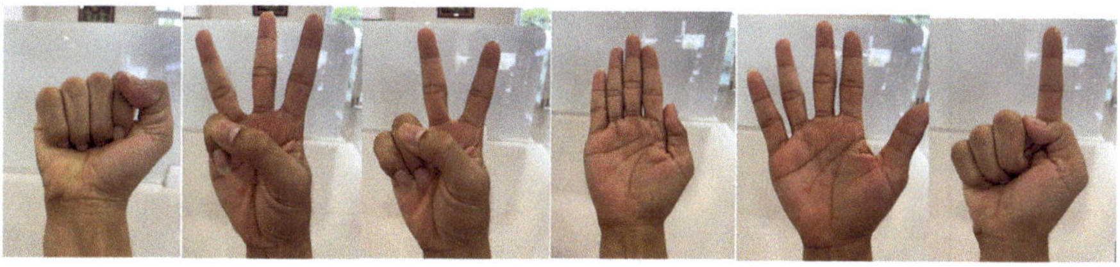

Fig. 2 The defined and trained hand postures

- Fist: A closed hand with all fingers curled inward, resembling a fist. This gesture can represent actions such as grabbing, selecting, or confirming.
- Open palm: An open hand with all fingers spread apart. This gesture can indicate a stop or halt command, an open-handed gesture for general interaction, or a hand position for receiving or giving objects.
- Thumbs-up: A hand gesture with the thumb extended upward and other fingers curled inward. This gesture often signifies approval, agreement, or a positive command.
- Thumbs-down: A hand gesture with the thumb extended downward and other fingers curled inward. This gesture typically represents disapproval, disagreement, or a negative command.
- Pointing: Extending the index finger outward to indicate a specific location or object. This gesture can be used for pointing at items on a screen, selecting options, or indicating direction.
- Peace sign: Extending the index and middle fingers outward while keeping the other fingers curled. This gesture often represents peace, victory, or a two-finger command.
- OK Sign: Forming a circle by touching the thumb and index finger while keeping the other fingers extended or relaxed. This gesture can convey approval, agreement, or a confirmation command.

It's important to note that the specific set of trained hand postures will depend on the requirements of the application or system. The system may be trained on a larger set of gestures, including more complex hand configurations or dynamic gestures involving hand movements. The training process involves collecting labeled data of the hand postures and using machine learning or pattern recognition techniques to learn the mapping between the hand postures and corresponding commands or actions shown in Fig. 3.

3 Experiment Results

Once the mobile device is connected to the MCU microcontroller, it sends the command to the home appliances after completion of hand gesture recognition and the execution time back to the computer. After testing, it takes 0.62 s from making a

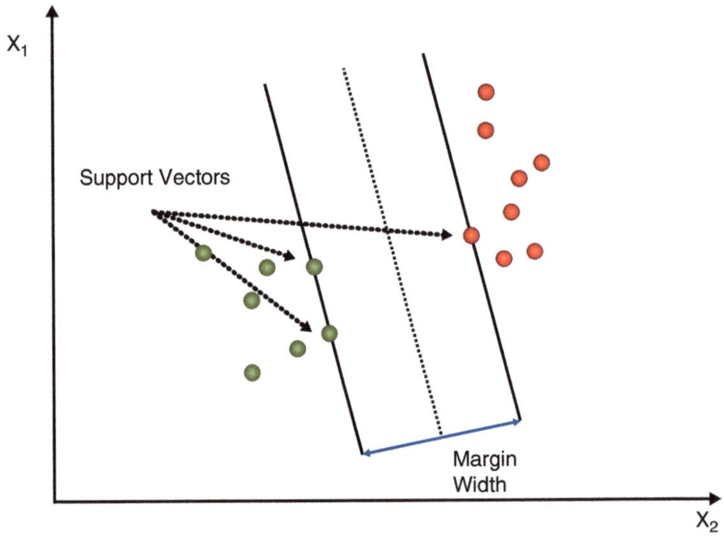

Fig. 3 SVM: support vectors representation with maximum-margin hyperplane

correct gesture to turn on the LED light. The original LED light is turned on, but when the finger makes the gesture of number 1, the command makes the LED light turn off, as shown in Fig. 4(a–d). Originally, the LED light turned off, when the finger makes the gesture of number 2, the command makes the LED light turn on, as shown in Fig. 4(e–h). Therefore, this test shows that it is possible to use gesture recognition to control the home appliances.

As shown in Fig. 4, there are eight controllable light beads on the LED light, and gestures can be used to make the light beads display according to the desirable brightness and color. As shown in Fig. 4h, when the finger makes the gesture of number 0, it sends a command to make the RGB light turn off. As shown in Fig. 4(a–h), when the finger makes the gesture of number 0-5, the command makes the LED light bar change the status on or off.

In this study, ten samples were asked to make number gestures that could be judged by the naked eye in bright light with a simple background. They performed ten defined gestures at different angles, with the palms facing outward or inward, in a total of five movement patterns.

4 Conclusion

This is study focused on the development of a hand gesture-controlled lighting system using an artificial intelligence algorithm. The proposed system aimed to interpret hand gestures such as waving, fist, and pinch to adjust the lighting

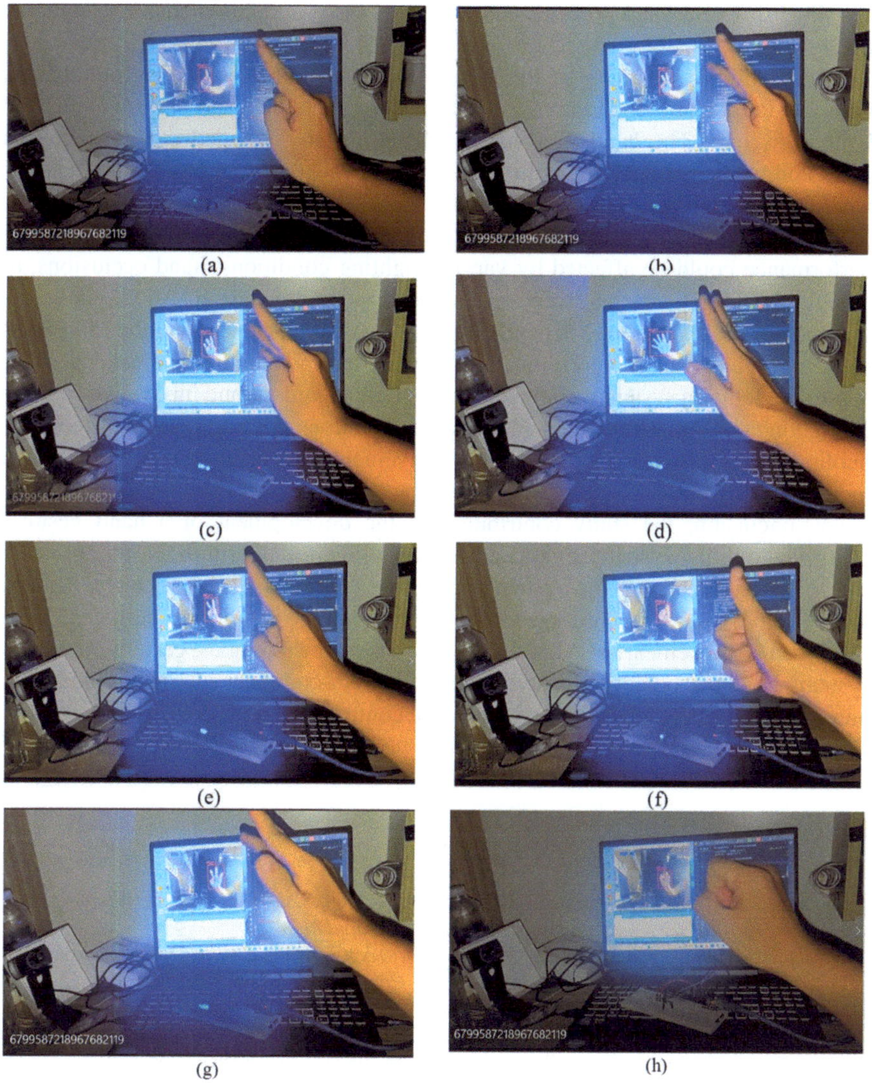

Fig. 4 Experiment result of hand gesture recognition

accordingly. The research involved the use of computer vision techniques, including hand detection, tracking, segmentation, and feature extraction, to capture and interpret hand movements.

The results of the study showed that the proposed system was feasible and effective in recognizing and interpreting hand gestures for controlling the lighting system. The combination of computer vision algorithms and artificial intelligence allowed for accurate and real-time gesture recognition. The system successfully classified different hand postures and translated them into commands for adjusting the lighting.

The usability evaluation of the system indicated that it was user-friendly and intuitive. Participants found it easy to learn and operate the hand gesture-controlled lighting system. The hands-free nature of the system provided convenience and accessibility, particularly for individuals with physical disabilities or in situations where traditional switches may be impractical. While the system demonstrated promising results, there were certain limitations and challenges. The system's performance could be affected by varying lighting conditions, hand occlusions, or complex hand gestures that were not included in the trained set. Additionally, the accuracy of the system may depend on the quality of the camera and the robustness of the computer vision algorithms. The commercial potential of the hand gesture-controlled lighting system was evident, as it offered a unique and innovative solution for lighting control. The system could find applications in smart homes, offices, and public spaces, providing users with a convenient and interactive way to adjust lighting settings.

In conclusion, the study contributed to the development of a hand gesture-controlled lighting system using an artificial intelligence algorithm. Further research can focus on enhancing the system's robustness, expanding the set of recognized hand gestures, and exploring additional applications in various domains. With continued advancements in AI and computer vision technologies, gesture-controlled systems have the potential to revolutionize human-computer interaction and offer new possibilities for intuitive and hands-free control of various devices and systems.

References

1. Di Donato, B., Dooley, J., Hockman, J., Bullock, J., & Hall, S. (2017). *MyoSpat: A hand-gesture controlled system for sound and light projections manipulation.*
2. Natarajan, K., Nguyen, T. H. D., & Mete, M. (2018, April). Hand gesture controlled drones: An open source library. In *In 2018 1st International Conference on Data Intelligence and Security (ICDIS)* (pp. 168–175). IEEE.
3. Bhuiyan, M., & Picking, R. (2009, November). Gesture-controlled user interfaces, what have we done and what's next. In Proceedings of the fifth collaborative research symposium on security, E-Learning, Internet and Networking (SEIN 2009), Darmstadt, Germany (pp. 26–27).
4. Sai Pravallika, G., Lakshmi Akhila, M., Divya, M., Madhu Babu, T., & Kranthi Kumar, G. (2022). Hand gesture controlled contactless elevator. In *Inventive communication and computational technologies: Proceedings of ICICCT 2021* (pp. 719–730). Springer.
5. Gourob, J. H., Raxit, S., & Hasan, A. (2021, July). A robotic hand: Controlled with vision based hand gesture recognition system. In *In 2021 International Conference on Automation, Control and Mechatronics for Industry 4.0 (ACMI)* (pp. 1–4). IEEE.
6. Jhamb, P., & Rehalia, A. (2020). Wireless hand gesture controlled multiple musical instruments. *International Journal of Information Technology, 12*, 19–25.
7. Tan, J., Shao, L., Lam, N. Y. K., Toomey, A., & Ge, L. (2022). Intelligent textiles: Designing a gesture-controlled illuminated textile based on computer vision. *Textile Research Journal, 92*(17–18), 3034–3048.

8. Gunapriya, B., Kumar, K. V., Zaidi, H. J., Reddy, L. N., Vikram, S., & Sadiq, J. (2022, February). A review of Arduino based hand gesture controlled robot using IoT. In *In 2022 Second International Conference on Artificial Intelligence and Smart Energy (ICAIS)* (pp. 1211–1214). IEEE.
9. Kang, S. P., & Katupitiya, J. (2004, September). A hand gesture controlled semi-autonomous wheelchair. In *2004 IEEE/RSJ International Conference on Intelligent Robots and Systems (IROS)(IEEE Cat. No. 04CH37566)* (Vol. 4, pp. 3565–3570). IEEE.

Occupational Safety From the Point of View of Safe Worker Behaviour in Peruvian Construction Companies

Carlos Huerta-Soto, Juan Villanueva-Calderón, Roger Rurush-Asencio, Jose Sifuentes-Stratti, and Hugo Maldonado-Leyva

Abstract The purpose of this study, for which 53 businesses were contacted, was to determine whether or not there is a correlation between employees' safe behavior on the job and their workplace safety. The need for this study, which seeks to identify the relationship of safe conduct with occupational safety in construction companies, arises from the rapid expansion of the construction industry in Peru, the high rate of construction-related accidents, and the lack of managerial concern for fostering a culture of safety in the workplace. The Kolmogorov-Smirnov normality test was carried out, and the result was a p-value less than 0.05. This indicates that neither of the variables has a normal distribution. The subsequent Spearman's Rho test yielded a value of 0.632, and the significance level was 0.00**. The study came to the following conclusion: there is some evidence to suggest that safe worker behavior and occupational safety in construction enterprises have a moderately positive association.

Keywords Accident prevention · Construction industry · Safe behavior · Workplace safety

1 Introduction

Occupational safety is a relevant issue in business management, and even more so occupational safety in companies in the construction sector. Nowadays, the occupational safety system is recognised as an approach to generate competitiveness in companies, besides being an essential element to prevent occupational accidents, it is

C. Huerta-Soto (✉) · J. Villanueva-Calderón
Universidad Cesar Vallejo, Lima, Peru
e-mail: chuertaso@ucvvirtual.edu.pe; vcalderonj@ucvvirtual.edu.pe

R. Rurush-Asencio · J. Sifuentes-Stratti · H. Maldonado-Leyva
Universidad Nacional Santiago Antunez de Mayolo, Huaraz, Peru
e-mail: rrurusha@unasam.edu.pe; jsifuentess@unasam.edu.pe; hmaldonadol@unasam.edu.pe

J. M. K. P. et al. (eds.), *Business Resilience and Digital Technology in the Post-Pandemic Era*, Contributions to Environmental Sciences & Innovative Business Technology, https://doi.org/10.1007/978-3-031-48075-1_4

necessary to examine the effect of workers' behaviour in construction companies. Workers in the construction sector are the most prone to work-related injuries, unlike other sectors, the construction sector is project-based, all different projects and each type of project (housing buildings, sports infrastructure and transport) has its own characteristics, which may include different working methods, materials used and techniques [1, 2].

In the Peruvian case, specifically in the Department of Ancash, the number of construction companies has grown over the last five years, due to the fact that the Ancash Region has become an area of mining influence due to the establishment, start-up and implementation of activities of two world-renowned mining companies such as: Minera Antamina S.A. and Barrick Gold Corporation. According to Corrales et al. [3], construction companies have a similar number of accidents as mining companies, despite the fact that the latter have better occupational safety conditions. On the other hand, the International Labour Organization [4] mentions that approximately 300 thousand accidents occur in construction companies, generated by poor occupational safety practices, one of the causes of occupational accidents are related to unsafe acts on the part of workers this is given by the few occupational safety measures. For Salvador [5] "accidents occur due to a variety of situations and among them are the factors derived from the distraction of the individual by worries, anxiety, lack of sleep or depression; which may be the result of an adverse work environment" (p. 36).

In Peru, [6] "Law on Safety and Health at Work". It establishes policies in the field of occupational safety, its application is recognised for all companies in both the production and service sectors, mentions the responsibilities of managers regarding the protection of their workers at all times, and also proposes the creation of Occupational Safety and Health Management Systems with representation of their collaborators. To this end, it modifies the current regulations and demands the achievement of a culture of occupational risk prevention, based on the obligatory nature of the duties regarding the prevention of occupational accidents, improving the state's supervisory function through the National Superintendence of Labour Inspection and the commitment of the workers' unions.

Haro et al. [7] point out that risk plans will not be successful if they lack support from the management of the organisations, and not only financially but must also provide moral support for the company's workers. Also Castejón & Benavides [8]; Ramirez-Asis et al. [9] point out that research related to occupational safety has discovered psychological indicators that added to the level of risk generate incidents and accidents at work, which is why it is necessary to explain the behaviours of workers in terms of safety; so it proposes to conduct the present research to determine the relationship of safe conduct with occupational safety, it is based on the observations made to employees of construction companies during the first half of 2019, we were encouraged to study these cases, because the impression that there were no safety standards, to which it was found that these standards did exist, then it was attributed to poor training and thus it was identified that the attitude of the worker was the main reason for incidents and accidents at work in this type of company.

According to Mendoza [10] the theory of behaviour at work sustains that in order for workers to carry out their work safely, they must have the following conditions: 1. to be able to work, 2. to know how to work, and 3. to want to work safely. Furthermore, in a study [11] states that "...the Behaviour-Based Safety methodology is one of the methodologies—but without doubt the most established, proven and effective available—to act on the third condition of the triconditional model" (p. 163).

The study pursues the following objectives: to determine the relationship of safe behaviour with occupational safety in construction companies, to identify the type of workers' behaviour and the level of occupational safety.

2 Methodology

A total of 523 workers from 53 construction companies in the department of Ancash, Peru were surveyed. The average age was 35 years and the standard deviation was 6.54; 86% of the sample was male and 14% female; 87% of the total number of workers were from the operative part and 13% from the administrative part; the average number of years of experience in the construction companies was 6 years. The technique used was simple random sampling. Workers with less than 6 months of seniority at the time of application were excluded.

To measure the safe behaviour variable, we turned to Aguilar et al. [12] who designed the safe behaviour scale, this questionnaire is proposed to measure the perception of workers about their own behaviours in the workplace, collaboration in safety conditions and the level of interest in the safety of the staff in their work area. This questionnaire has 14 questions with Likert-type alternatives from 1 to 5. In addition, this questionnaire was tested for content validity by three experts and obtained a Cronbach's alpha value of 0.89.

To measure occupational safety, the questionnaire designed by Gómez et al. [13] was used and they propose 7 dimensions, 26 groups and 79 indicators, for the present study an adaptation was made, finally remaining with four dimensions: The first dimension, protection at work which consists of 17 items; the second dimension, ergonomics and consists of 4 items, the third intralaboral psychosocial factors which consist of 8 items and finally extralaboral psychosocial factors which consist of 6 items. The final questionnaire consists of 35 items, the content was validated by the review of three experts and obtained a Cronbach's alpha value of 0.92.

The present research was exploratory, descriptive and correlational [14]. The questionnaires were carried out individually in the workplace, and then entered into a database in the SPSS V24 programme. Descriptive statistics and cross-tabulations were used to determine worker behaviour and occupational safety; to test the hypothesis, the Kolmogorov-Smirnov normality test was performed, obtaining a $p < 0.05$, which shows that both variables do not have a normal distribution, so to identify the relationship between the two variables, Spearman's Rho statistic was used.

3 Results

From the table above it can be seen that behaviour varies according to the level of education of the worker, as the level of education increases, behaviour improves (Table 1).

Figure 1 shows that more than 57.5% of workers in construction companies perceive negative safety behaviour (Table 2).

Workplace safety improves as the size of a company gets larger, because its workplace safety procedures and conditions are more standardised (Fig. 2).

65.2% of workers in construction companies perceive workplace safety as negative (Fig. 3).

46.5% of construction workers perceive a secondary level of accident prevention at work (Table 3).

39% of workers in construction companies perceive that behaviour is indifferent and the level of accident prevention is secondary (Table 4).

Table 1 Safe behaviour according to educational level

Level of education	Safe behaviour			
	Positive	Medium	Negative	Total
Secondary school incomplete	5.56%	1.96%	12.87%	20.39%
Secondary	9.49%	5.02%	37.95%	52.45%
Higher technical	9.92%	2.62%	4.14%	16.68%
University tertiary	6.76%	1.20%	2.51%	10.47%
Total	31.73%	10.80%	57.47%	100.00%

Fig. 1 Levels of safe behaviour

Table 2 Safety at work according to company size

Company size	Workplace safety			
	Positive	Medium	Negative	Total
Micro enterprise	9.38%	1.85%	5.56%	16.79%
Medium enterprise	10.58%	4.80%	39.91%	55.29%
Large Enterprise	3.27%	2.73%	21.92%	27.92%
Total	23.23%	9.38%	67.39%	100.00%

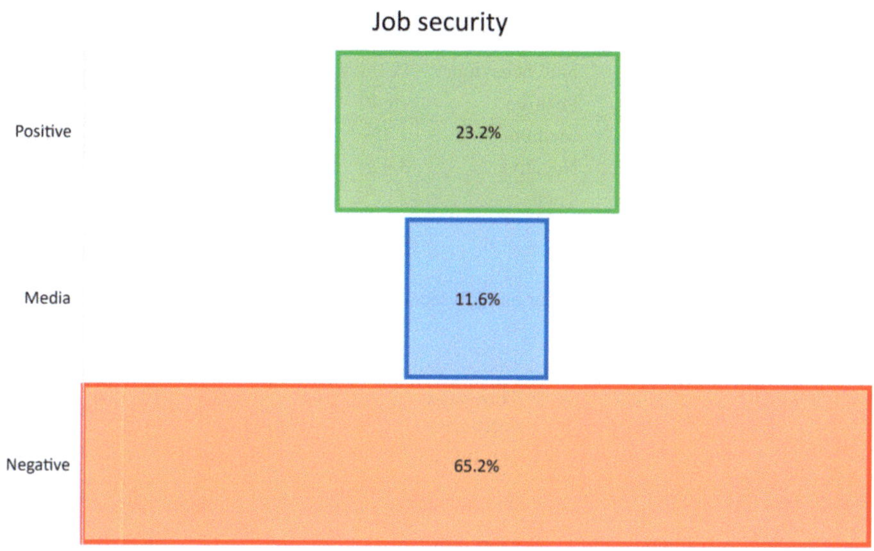

Fig. 2 Levels of workplace safety

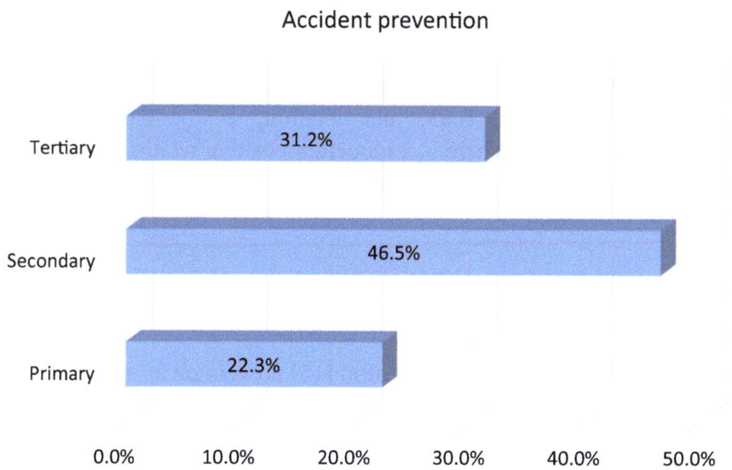

Fig. 3 Levels of accident prevention

Table 3 Accident prevention according to behaviour

Behaviour	Prevention of accidents at work			Total
	Primary	Secondary	Tertiary	
Assertive Behaviour	9%	5%	2%	16%
Indifferent Behaviour	10%	39%	6%	56%
Passive Behaviour	3%	2%	23%	28%
Total	22%	47%	31%	100%

Table 4 Safety at work according to Safe behaviour

Safe behaviour	Workplace safety			
	Positive	Medium	Negative	Total
Positive	18.4%	2.1%	11.2%	31.7%
Medium	1.4%	5.2%	4.1%	10.8%
Negative	3.4%	4.3%	49.8%	57.5%
Total	23.2%	11.6%	65.2%	100.0%

Table 5 Correlation of safe behaviour and job security

			Safe behaviour	Workplace safety
Rho de Spearman	Safe behaviour	Correlation coefficient	1.000	0.632**
		Sig. (bilateral)		0.000
		N	523	523
	Workplace safety	Correlation coefficient	0.632**	1.000
		Sig. (bilateral)	0.000	
		N	523	523

**. La correlación es significativa en el nivel 0.01 (2 colas)

49.8% of workers in construction companies perceive work safety as negative, due to negative safe behaviour (Table 5).

Using Spearman's Rho = 0.632 and obtaining a $p < 0.01$**, the research hypothesis is demonstrated: There is a moderate positive relationship between safe behaviour and occupational safety in construction companies.

4 Discussion

According to the World Health Organisation [15] prevention at work occurs at three levels: primary, secondary and tertiary, which propose differentiated objectives and techniques, in the present research it was shown that 47% of workers in construction companies perceive a level of secondary prevention based on an indifferent behaviour with respect to accidents at work, similar result was shown by Rodrigues et al. [16] who did a study on occupational risks in Colombian companies and showed that secondary prevention activities are developed through a rigorous entrance examination.

It is necessary to consider the participation of all workers in the execution of occupational safety activities in their workplace, since it is from these interactions that the perception of safety conditions is generated and supported by a type of work behaviour [17, 18]. Now, with respect to safe conduct 57% of workers in construction companies perceive negative safe conduct, this differs greatly with Nunu et al. [19] who, in 40 workers out of 244, in a cement factory, used a questionnaire that helps to measure the attitude of the worker, resulting that 60% of the respondents had positive safe conduct.

Aguilar et al. [12] mention that the perception of workers regarding health and safety at work is good, in addition to the fact that the bosses always use their personal protective equipment and show great interest in the safety at work of their workers. These results differ from the findings of the present research where the perception of safety at work is negative, which is based on the context, due to the accelerated and disorderly growth of construction companies in the Ancash region of Peru.

According to Thommandru et al. [20], Argüello-López et al. [21] the worker's behaviour towards risks of the employees of the operative part resulted to be in a medium-low level, this is an important aspect, since the workers of the operative part have influence of safety and health at work [22] this result is similar to those found in the present research, where the workers of the construction companies perceive a negative safe behaviour.

Other research has been published on attitudes and occupational accidents in manufacturing companies [23] and the study of attitudes towards occupational hazards in three SMEs in the construction sector [21, 24], the present research is an important contribution that seeks the relationship between safe behaviour and occupational safety in construction companies.

5 Conclusions

The low predisposition of the workers due to their level of education represents a limitation for the present study, together with the location of the Ancash region in the highlands of Peru; in order to generalize the results to all construction companies in Peru, it would be necessary to carry out research in other regions.

In conclusion, the reason for this research was to demonstrate the existence of a moderate positive correlation between safe behavior and occupational safety in construction companies, due to the negative safe behavior and the negative perception of occupational safety in construction companies in the Ancash region.

References

1. Shaikh, A. A., Lakshmi, K. S., Tongkachok, K., Alanya-Beltran, J., Ramirez-Asis, E., & Perez-Falcon, J. (2022). Empirical analysis in analysing the major factors of machine learning in enhancing the e-business through structural equation modelling (SEM) approach. *International Journal of System Assurance Engineering and Management, 13*(Suppl 1), 681–689. https://doi.org/10.1007/s13198-021-01590-1
2. Yoon, S. J., Lin, H. K., Chen, G., Yi, S., Choi, J., & Rui, Z. (2013). Effect of occupational health and safety management system on work-related accident rate and differences of occupational health and safety management system awareness between managers in South Korea's construction industry. *Safety and Health at Work, 4*(4), 201–209. https://doi.org/10.1016/j.shaw.2013.10.002
3. Corrales, C., Sánchez, C., & Toledo, G. (2014). Estudio de la Siniestralidad en Seguridad en Empresas del Sector Construcción a partir de la Nueva Legislación Peruana. In *12th Latin American and Caribbean Conference for Engineering and Technology 1–11.* Guayaquil. https://bit.ly/33yNiyG
4. International Labour Organization (14 de 03 de 2019). *Organización Internacional de Trabajo.* https://bit.ly/39bpB0J
5. Salvador, M. J. E. (2018). Riesgos psicosociales del aeropuerto de Manta. *Revista San Gregorio, 22*, 30–39. https://doi.org/10.36097/rsan.v1i22.610
6. Ley N° 29783 (2011). Ley de seguridad y salud en el Trabajo. Diario oficial el peruano. 20 de agosto del 2011.
7. Haro, N. I. G., Hernández, J. G. V., Dávalos, J., & Torres, M. D. L. M. (2020). Integración de la seguridad y salud laboral y la Responsabilidad Social Empresarial en una embotelladora de México: Un estudio de caso. *Hitos de Ciencias Económico Administrativas, 26*(75), 197–212. https://doi.org/10.19136/hitos.a26n75.3990
8. Castejón, E., & Benavides, F. G. (2021). De la seguridad e higiene a la seguridad y salud: 50 años de prevención de riesgos laborales en España. *Archivos de Prevención de Riesgos Laborales, 24*(2), 99–103. https://doi.org/10.12961/aprl.2021.24.02.02
9. Ramirez-Asis, E., Maguina, M. E., Infantes, S. E., & Naranjo-Toro, M. (2020). Emotional intelligence, competencies and performance of the university professor: Using the SEM-PLS partial least squares technique. *Teaching Innovation in the Development of Professional, 23*, 99–114. https://doi.org/10.6018/reifop.428261
10. Mendoza, L. (2019). Gestión de la seguridad basada en comportamientos. *Revista San Gregorio, n.31*, 138–149. https://doi.org/10.36097/rsan.v0i31.964
11. Torres-Sandoval, F. A. (2019). Propuesta de seguridad basada en el comportamiento para una empresa de transporte público en Colombia. Continuación de un caso de estudio. *Dyna, 86*(209), 378–387. https://doi.org/10.15446/dyna.v86n209.73816
12. Aguilar, C., De Lille, M., Escamilla, M., & Cetina, T. (2018). Clima de Seguridad Ocupacional, Respuesta de Seguridad del jefe y Conducta Segura del Trabajador. *Revista Electrónica de Psicología Iztacala., 21*(1), 121–141. https://www.medigraphic.com/cgi-bin/new/resumen.cgi?IDARTICULO=81516
13. Gómez, A., Silva, M., Vilaret, A., & Suasnavas, P. (2016). Diseño metodológico de la encuesta sobre condiciones de seguridad y salud en el trabajo en Ecuador. *Innova Research Journal, 1*(8), 1–14. https://doi.org/10.33890/innova.v1.n8.2016.41
14. Hernández-Sampieri, R., & Mendoza, C. (2018). Metodología de la investigación. In *Las rutas cuantitativa, cualitativa y mixta* (p. 714p). Editorial Mc Graw Hill Education. isbn:978-1-4562-6096-5.
15. World Health Organisation, WHO (12 de 08 de 2017). Protección de la salud de los trabajadores. https://bit.ly/39dVKV8
16. Rodrigues, F., Antunes, F., & Matos, R. (2021). Safety plugins for risks prevention through design resourcing BIM. *Construction Innovation, 21*(2), 244–258. https://doi.org/10.1108/CI-12-2019-0147

17. Pallathadka, H., Ramirez-Asis, E. H., Loli-Poma, T. P., Kaliyaperumal, K., Ventayen, R. J. M., & Naved, M. (2021). Applications of artificial intelligence in business management, e-commerce and finance. *Materials Today: Proceedings*. https://doi.org/10.1016/j.matpr.2021.06.419
18. Thill, F. (2022). Information Security Risk Management. In *Privacy and Identity Management. Between Data Protection and Security: 16th IFIP WG 9.2, 9.6/11.7, 11.6/SIG 9.2. 2 International Summer School, Privacy and Identity 2021, Virtual Event, August 16–20, 2021, Revised Selected Papers* (pp. 17–22). Springer International Publishing. https://doi.org/10.1007/978-3-030-99100-5_2
19. Nunu, W. N., Kativhu, T., & Moyo, P. (2018). An evaluation of the effectiveness of the behaviour based safety initiative card system at a cement manufacturing company in Zimbabwe. *Safety and Health at Work, 9*(3), 308–313. https://doi.org/10.1016/j.shaw.2017.09.002
20. Thommandru, A., Espinoza-Maguiña, M., Ramirez-Asis, E., Ray, S., Naved, M., & Guzman-Avalos, M. (2021). Role of tourism and hospitality business in economic development. *Materials Today: Proceedings*. https://doi.org/10.1016/j.matpr.2021.07.059
21. Argüello-López, G., Uribe-Bermúdez, J., & Valdivieso-Guerrero, M. (2017). Relación entre capacitación y actitud hacia los riesgos laborales en el sector construcción del área metropolitana de Bucaramanga. *I+ D Revista De Investigaciones, 9*(1), 14–26. https://doi.org/10.33304/revinv.v09n1-2017002
22. Kouabenan, R., Ngueutsa, R., & Mbaye, S. (2015). Safety climate, perceived risk, and involvement in safety management. *Safety Science, 77*, 72–79. https://doi.org/10.1016/j.ssci.2015.03.009
23. Taufek, B., Zulkifle, B., & Kadir, B. (2016). Safety and health practices and injury management in manufacturing industry. *Procedia Economics and Finance, 35*, 705–712. https://doi.org/10.1016/s2212-5671(16)00088-5
24. Yating, Y., Mughal, N., Wen, J., Ngan, T. T., Ramirez-Asis, E., & Maneengam, A. (2022). Economic performance and natural resources commodity prices volatility: Evidence from global data. *Resources Policy, 78*, 102879. https://doi.org/10.1016/j.resourpol.2022.102879

Strategic HRM and Global Recession: Emerging Challenges and Opportunities in the Digital Era

Alamelu Mangai Raman and Jaffer Ali Khan

Abstract Recessions have been a constant feature of contemporary economic history as the global economic downturn affects the majority of nations at the moment. Human resource management (HRM) has been further tested by the global economic downturn and the contraction of enterprises, necessitating a rethinking of how organizations interact with their workers in terms of value addition, staff reduction, skill enhancement, and the introduction of multitasking. Companies that are struggling to stay afloat in the face of the global economic slump are not likely to invest much on their human resources. The economic crisis may be a bad time for many, but for firms with solid financial footing and attractive business models, it provides crucial chances to enhance their human resource management skills and prepare themselves for the eventual comeback. Governments, the commercial sector, and labor unions in developing nations face unique problems from the current form of globalization, which is biased toward wealthy countries. "Meltdown" and "downturn" are the result of Western economies' insatiable need for growth and their stubborn refusal to let emerging nations have a say in establishing the international economic framework. Hence, this research takes a strategic approach to HRM in order to tackle these one-of-a-kind problems.

Keywords Modern economics · Economic recession · Human resource management · Globalization · Strategic approach

1 Introduction

Countries throughout the globe are working jointly in search of a mutually agreeable approach to solving the global economic crisis so that everyone's economy may be saved. Governments, the commercial sector, and labor unions in developing nations

A. M. Raman (✉) · J. A. Khan
College of Economics and Business Administration, University of Technology and Applied Sciences-Al Musannah, Musannah, Sultanate of Oman
e-mail: Alamelu@act.edu.om; Jaffer@act.edu.om

J. M. K. P. et al. (eds.), *Business Resilience and Digital Technology in the Post-Pandemic Era*, Contributions to Environmental Sciences & Innovative Business Technology, https://doi.org/10.1007/978-3-031-48075-1_5

face unique problems from the current form of globalization, which is biased toward wealthy countries. “Meltdown” and “downturn” are the result of Western nations” insatiable need for growth and their stubborn refusal to let emerging nations have a say in establishing the international economic framework. Because of this, businesses in emerging markets are also feeling the pinch. Organizations (both domestic and foreign) are crucial market drivers that are being hit hard by the current recession; as such, their input into the search for a way out of the current impasse is essential. The focus of this chapter is not on the group’s overall strategy but rather on the specific steps each member of the group should take, the aspects of their company that need the most attention right now, and the ways in which they may tailor their approach to these challenges. Unquestionably, a recession is a crisis. Modern management literature advises over and again that “out-of-the-box” thinking is required of organizations during times of crisis. The rule of thumb does not always apply.

Companies may be creative if they constrain their thinking to predefined parameters, sounds like the train of thought and structure institutions need to construct in order to make the most of their existing assets. Human resources (HR) provide an extra layer of assistance during a crisis, setting them apart from other resources. Human resources now have their chance to shine. The connection among human resources and the tactical goals of the organization needs to be improved to guarantee survivability (immediate objective) and accomplishment (long-term priority) during the current crisis. The plan addresses SHRM, or strategic human resource management. Although this technique may seem antiquated, it has the potential to save the company beyond every other tactic the company has developed so far.

1.1 Global Recession: An Outsider View

According to the NBER (US National Bureau of Economic Research), a recession is “a broad and deep contraction of economic activity that persists for longer than a few month.” Individuals in the Americas are now experiencing a downturn, yet economic downturns are not unique to America. There has been a recession in every nation with even a somewhat comparable contemporary economic system. Recession hit the United States for the first time between 1797 and 1800. The panic of 1797 was triggered in large part by the Bank of England’s deflationary policies making their way over the Atlantic Ocean to the United States. The business-oriented real estate market in the United States was shaken up by this. The economy of Britain had been struggling before it became involved in the conflicts against France that broke out during the French Revolution. This is only one illustration of how swiftly a country’s economic downturn may ripple across borders. Economists are in agreement that what happens in one nation, particularly an important one, will have repercussions around the globe. Since the Panic of 1797, the US economy has gone through several recessions due to a variety of tribulations. The good news is that the Panic of 1797 was the final economic downturn of the 1800s.

The next proven period of economic contraction took place between 1807 and 1814; this period is known as the Recession of 1807. The Federalist Party battled hard against this measure, which led to the rise of trafficking in New England and the destruction of several shipping-related companies. Then came the 1819 Panic. For the young economy of the United States, this was the first serious financial crisis. It produced extensive foreclosures, bank failures, substantial joblessness, and a precipitous decline in industry and agriculture. The immense economic growth that had ensued in the wake of the War of 1812 ended with this downturn. The Great Depression of 1929–1932 had far-reaching effects on the international economy. The true causes of the Great Depression were the collapse of the real estate market and the unthinking adoption of the Anglo-Saxon economic paradigm.

Recession has hit several nations in 2009 for identical causes. This is a major reason why the global economy is struggling. There has been much discussion of a "global recession" among influential economic groups and corporate executives. The current slump is a result of Western governments' and corporations' excessive desire for profit. The current crisis has not only shown the world the dark side of capitalism, but it has also shown that "capitalism" has become "exhausted capitalism."

1.2 Strategic HRM: A Brief Outlook

A sufficient indication of the curiosity and course of the business community is the amount of attention and significance being accorded to the strategic management of human resources (hereafter SHRM) not only in dissertations and the work made by management specialists but also in commercial board debates. The persistence of SHRM is to increase organizational performance by the effectiveness of HR practices, policies, and procedures with business tactic. For an organization to practice SHRM, it must see the HR department as an integral element of strategy development and execution via HR processes including hiring, training, and compensation. Strategic HR, on the other hand, acknowledges HR's partnering role.

Human resources strategies, within the context of business planning, are the predetermined HR actions the organization will take to accomplish its goals. According to the literature [1–3], SHRM is the process of managing a company's human capital (its personnel) so that they help the company achieve its commercial goals. Schuler [4, p. 18] provides a more wide-ranging speculative description of SHRM. It is all about putting the parts together and adjusting them as needed. Its emphasis is on ensuring that (1) line managers and staff incorporate HR practices into their work on a regular basis; (2) HR policies are consistent across policy fields and organizational levels; and (3) HR management is completely consistent with the officialdom's tactics and strategic objectives.

Strategic human resource management (SHRM) is "the sequence of scheduled human resource investments and operations aimed at assisting an organisation in achieving its goals" (p. 298), according to Wright and McMahan [5]. Some of the most often mentioned foundational SHRM features are as follows: Human resources are seen as an investment rather than a cost; SHRM practices are connected to improved organizational performance; and a competitive advantage is achieved via a focus on aligning HR and business objectives.

2 Theoretical Foundations of Strategic HRM

Many theoretical standpoints have been formed to structure our understanding of how strategic issues affect HR practices, as will be briefly addressed here. Wright and McMahan's [5] six theoretical influences provide the foundation of a comprehensive theoretical framework. Four of these factors explain strategic concerns-motivated actions. Two examples are the company's resource-based approach and its behavioral viewpoint. Human resource activities that are not driven by business strategy can be explained using resource dependency and institutional theory. The company's resource-based paradigm combines legislative finances with strategic leadership [6]. According to this school of thought, one of the major variables determining a company's competitive edge is its resources. Companies can establish a modest lead only by providing worth in a move that rivals find grim to replicate.

Value may be created by the use of conventional competitive advantages, such as economic and physical assets, technological advances, and economies of scale. The counterargument, based on the availability of these resources, is that they are becoming easier to obtain and copy. Consequently, they are less important for gaining an edge in business competition, particularly when compared to a more intricate social structure like an employment system. According to the Research of Jackson and Schuler [7] and Pfeffer [8], human resource strategies and applies may be a significant basis of a company's Over time, you will gain a competitive advantage. According to Barney [6], four empirical metrics of a firm's ability to develop a competitive advantage from its resources are value, rarity, imitability, and substitutability. That is, for a corporation to have a competitive advantage, it needs resources that are both unusual and not commonly available. Physical resources include an organization's physical plant, technological and equipment, and geographic location; human resources include its experienced and knowledgeable employees; and organizational resources include its structure; systems for planning, monitoring, and directing actions; and communal relationships within and between the group and external constituencies. To get an edge over the competition, businesses may employ HR practices to improve their people and organizational resources [9].

Another conceptual factor is the contingency-based behavioral perspective. This perspective examines methods used to manipulate beliefs and actions and highlights the instrumentality of these methods in accomplishing strategic goals. The homeostatic system illustrates how feedback on HR practices' achievements to strategies leads to their continuation or discontinuation. As an example, training programs might be implemented to better execute a plan and then modified or scrapped as necessary. The usage of control mechanisms like performance assessment and incentive systems is justified by the fourth factor, transaction costs. The idea here is that methodologies may be abandoned if performance evaluation systems are not linked to incentive systems. Neither the resource-based nor the institutional theories are motivated by strategic concerns; instead, they explain HR practices in terms of power and politics, resource oversight, and ethical obligations [10, pp. 107–108].

3 A Response to Challenges of Global Downturn

HR experts may have a strategic impact during this economic downturn by stepping in and taking charge. The macroeconomic setting is the starting point for analysis in the traditional strategy paradigm. Then, we take a peek at the local setting and see how it has an effect on ourselves and our rivals. The next step is to identify the strategic elements over which HR has direct control. We resort to dirty tricks at last. A human resources department that can think on its feet during a recession is essential. The HRM Department is being tasked with introducing novel concepts, revamping HRM processes, and developing new ones. In addition, the effort must not break the bank or prevent the company from losing money. Only if the company has a strict policy on SHRM can this be achieved.

Although it is theoretically claimed (but not clearly true) that the recession is not suitable for large-scale innovative practices, but which are rather straightforward during periods of economic expansion, the proper use of SHRM techniques may disprove these statements. During a recession, human resources must show a decisive protagonist in order to save the company money. When the next growth age arrives, the creative ideas and solutions that were made possible by all of the assistance functions will make the organization stronger and more cost-effective. During a recession, HR management should rely on the following tried-and-true, adaptable, and widely accepted strategic human resource management (SHRM) formulations:

- To maximize manpower strength.
- Taking deliberate steps to boost the overall productivity and efficiency of the organization.
- Working on compensation benefits.
- Rethink your training and development programs.
- Ensure that the policies and handbooks of the organization are up to date.
- Layoffs must be the final resort.
- Alternative workweek schedules and flexible scheduling might help firms increase output while decreasing overhead expenditures.
- Pay reduction must also be addressed at the final level.

Jobs can be changed, and responsibilities might be expanded to reflect the changes. In addition, HR departments must come up with creative responses to the economic downturn, such as recognizing their really important workers and bringing them into the inner circle of decision-makers. In a period of economic uncertainty, HR management must focus on the long-term effects of HRM innovations and set priorities accordingly. SHR management's purpose is not short-term cost-cutting but rather the fortification of the business in preparation for its anticipated expansion.

In times of economic hardship, SHRM may be an invaluable resource for overcoming difficulties from the employer's perspective.

1. Top management should be aware of the backup plan.
2. Hold regular brainstorming sessions with top management, and consider their proposals for strategic planning.
3. A clear message to employees that the organization will continue to attract critical personnel even in difficult circumstances.
4. Examine employee performance reviews to identify important personnel that the organization cannot afford to lose.
5. Communication should flow from top to bottom to create a conducive environment within the organization.
6. Constant assessment of all HR policies, processes, and procedures to verify that they are meaningful and directly contribute to the company's performance.
7. Firm advise to managers on dealing with the change management process.

4 Conclusion

This chapter demonstrated that SHRM is a remedy for economic downturn. Organizations may slow the erosion that occurs as a natural by-product of a downturn if they formulate their plans collaboratively and consistently put them into action. HRM is more challenging in tough economic times than in periods of fast expansion. Hence, HR has a vital part to play in preserving harmony at all levels of the organization and implementing changes that are both effective and well received in response to external pressures.

References

1. Baird, L., & Meshoulam, I. (1988). Managing the two fits of strategic human resource management. *Academy of Management Journal, 13*, 116–128.
2. Delery, J., & Doty, D. (1996). Modes of theorizing in strategic human resource management: Tests of universalistic and performance predictions. *Academy of Management Journal, 39*, 802–835.
3. Huseli, M. A., Jackson, S. E., & Schuler, R. S. (1997). Technical and strategic human resource management effectiveness as determinants of firm performance. *Academy of Management Journal, 40*(1), 171–188.
4. Schuler, R. S. (1992). Strategic human resource management: Linking people with the needs of business. *Organizational Dynamics, 21*, 18–32. http://dx.doi.org/10.1016/0090-2616(92)90083-Y

5. Wright, P., & McMahan, G. (1992). Theoretical perspectives for strategic human resource management. *Journal of Management, 18*, 295–320.
6. Barney, J. (1991). Firm resources and sustained competitive advantage. *Journal of Management, 17*, 99–120.
7. Jackson, S. E., & Schuler, R. S. (1995). Understanding human resource management in the context of organizations and their environments. *Annual Review of Psychology, 46*(1), 237–264.
8. Pfeffer, J. (1994). *Competitive advantage through people unleashing the power of the workforce*. Harvard Business School Press. Boston.
9. Schüler, R. S., & MacMillan, I. C. (1984). Gaining competitive advantage through human resource management practices. *Human Resource Management, 23*(3), 241–255. https://doi.org/10.1002/hrm.3930230304
10. Greer, C. (1995). *Strategy and human resources: A general managerial perspective*. Prentice Hall.
11. Barney, J., & Wright, P. (1988). On becoming a strategic partner: The role of human resources in gaining competitive advantage. *Human Resource Management, 37*, 31–46.

Service Quality Evaluation of Healthcare Sustainability: Patient Satisfaction, Service Delivery, and Clinical Outcomes

Durai Ravindran, Rajani Hariharan Pillai, Deeksha Srinivasa, and Sunitha Somashekaraiah

Abstract Since the definition of quality healthcare varies among consumers, experts, managers, policymakers, and payers, the main objective of this study is to provide a description of good healthcare that reflects the needs and expectations of all relevant stakeholders. This research aims to experimentally understand the quality of healthcare in Iran. Key healthcare stakeholders were interviewed individually and in focus groups. The definition of quality healthcare is providing effective and efficient healthcare services that meet the demands of both patients and providers while adhering to the latest professional principles and standards. All stakeholders agree that high-quality healthcare involves delivering efficient care that enhances patients' well-being and satisfaction. This study highlights the complexity of excellent healthcare, furthering our understanding of it. The findings have direct implications for medical practitioners who are encouraged to regularly assess the quality of healthcare using the identified characteristics. They can implement programs for continuous quality improvement to ensure high levels of patient satisfaction. For the first time, a comprehensive description of healthcare provision has been provided, incorporating the opinions and demands of various healthcare stakeholders. With the support of improved quality and efficiency in the entire healthcare environment, providers will be able to achieve reliable, affordable, and sustainable healthcare processes, as well as their goals of enhancing care delivery and patient outcomes. A company's service quality is determined by how effectively it meets the expectations of its clients. Customers purchase services based on specific

D. Ravindran (✉)
School of Management, Kristu Jayanti College (Autonomous), Bengaluru, Karnataka, India

R. H. Pillai
School of Commerce, Mount Carmel College, Autonomous, Bengaluru, Karnataka, India
e-mail: rajani.h.pillai@mccblr.edu.in

D. Srinivasa
Bengaluru City University, Bengaluru, Karnataka, India

S. Somashekaraiah
Department of Commerce, New Horizon College, Bengaluru, Karnataka, India

J. M. K. P. et al. (eds.), *Business Resilience and Digital Technology in the Post-Pandemic Era*, Contributions to Environmental Sciences & Innovative Business Technology, https://doi.org/10.1007/978-3-031-48075-1_6

requirements. The goal of total quality management in the healthcare industry is to improve patient outcomes while assessing the health benefits of hospitals and physicians' work. Quality management in the healthcare sector aims to reduce errors and enhance patient care. IBM developed the statistical software package known as SPSS Statistics, which includes features for data management, analytic tools, multimodal analytics, data analytics, and police cases. Spa Inc. existed for some time before being acquired by IBM in 2009. The latest versions are sold under the brand name "IBM SPSS Statistics." Inpatient, outpatient, relative, medical doctor, provider, manager, policymaker, supplier, quality manager, and accreditation staff are all considered in the outcome of Cronbach's alpha reliability. The total Cronbach's alpha score of the model is 0.590, indicating a reliability level of 33%. The aforementioned 373% increase in Cronbach's alpha product from the review article can be taken into consideration for analysis. Healthcare services, medical doctors, and inpatients are key components of the study. According to the latest clinical recommendations and standards, quality healthcare is described as consistently pleasing the patient by offering efficacious, effective, and efficient healthcare services that meet the patient's demands and satisfy providers. All stakeholders' conceptions of healthcare quality include the provision of effective care that enhances patient satisfaction and well-being.

Keywords Service quality · Healthcare · Sustainability

1 Introduction

In many ways, a hospital's ability to adjust to organizational, operational, and technological changes determines its level of excellence. It is crucial to take into account the growing patient population when constructing a hospital, as well as the long-term efficacy, efficiency, and adaptability of its design. Each of these factors contributes to high-quality healthcare and serves as a guide for the management process [1]. Additionally, most of the models used today are European in nature and have not been modified to suit the cultural and economic circumstances of developing nations. Measures such as general service performance, which have been developed in previous studies, may not be entirely adequate in determining the perceived quality of care. Moreover, previous studies have been overly brief, placing too little emphasis on technical elements and focusing solely on the functional aspect of services. These limitations allow for a margin of error. Thus, it is suggested that healthcare organizations develop better internal metrics to gauge the quality of their services. It has been argued that there is an ongoing debate about the standard of healthcare. Questions regarding the quality of the healthcare system and how different stakeholders perceive it have persisted for years. There is a general agreement that health quality is a complex, multidimensional concept that cannot be uniformly defined [2]. Negative perceptions of health service quality have adverse effects on the sector whenever service delivery is compromised. The loss of customers due to subpar service is analyzed to assess this impact. A model for prospective client loss is proposed, which is based on probabilistic hypotheses about how customers will act when they receive subpar service. This approach gives an economic boost to service quality and cannot be accomplished without a strong relationship between managerial emphasis and customer satisfaction outcomes. To compete and manage your finances, a three-pillar strategy for company healthcare provision is suggested in this chapter. The improved health services model (EHSM) refers to these three pillars. It argues that the health industry should adopt a systematic approach to providing healthcare, considering quality, cost, and performance aspects rather than focusing on improving only one aspect in isolation [3]. Studies were selected if they met the following criteria: they measured aspects related to the models, techniques, and tools used to assess the quality of hospitals or healthcare services; they discussed the use of measurement tools, samples, or instruments; or they referred to medical care or the hospital environment. Only original research, case studies, and documentation were included. Ad hoc strategies were employed to select appropriate and relevant studies from the reference list. After completion, researchers removed duplicate articles with similar topics from all sources. Documents were screened using predetermined content and exclusion criteria [4]. Due to global competition in the healthcare sector, patients are becoming more interested in and concerned about the delivery of healthcare services. The focus on health will continue to grow as the senior population rises, which has significantly increased health demands and altered healthy habits in the overall community. Residences and businesses, particularly health services, have been affected by recent expansions in the highly competitive global service sector. There is a correlation between health and these expansions. With the increasing need for high-quality healthcare, this emphasis has begun. Customers are compelled to choose the best hospital due to competition among hospitals [5]. There is considerable interest in healthcare quality. Finding healthcare systems that satisfy multiple stakeholders remains a challenge for practitioners and educators. The healthcare industry is under strain, and legislation intervenes by emphasizing patients' rights to quality care in order to lower costs and relieve strain on the healthcare system. Consequently, the challenge lies in how to maximize profits by offering patients high-quality service within available budgets in a competitive marketplace. The response from many nations is awaited. Japan serves as a standard example. Japanese healthcare professionals are under pressure from political influence and strong insurance corporations to reduce costs at the expense of patient care. Such circumstances have a significant impact on clinical [6].

2 Materials and Concepts

2.1 Inpatient

Patients who have been hospitalized are given inpatient treatment. Modern medicine has made significant strides, and the establishment of comprehensive outpatient clinics ensures that patients are admitted to hospitals only when they are critically ill or injured. Planning the patient's discharge from the hospital is frequently carried out by healthcare providers specializing in rehabilitation. Considerations for patient discharge include the patient's current health status, place of residence, and the type of support they can access. Even if the patient is deemed suitable for discharge, it is crucial to take into account factors such as the risk of re-injury while considering the patient's current predicament, to reduce healthcare expenses [7].

2.2 Outpatient

Effective care and prompt treatment are crucial for the patient's speedy recovery. An essential first step in any medical care is the outpatient department (OPD), which aids in the patient's recovery from or improvement of a certain physical ailment. The OPD is a section that offers additional services such as diagnosis, lab work, and preventive counseling. It promotes actions like immunization and regulates the transmission of infectious diseases. Despite the many useful features, quality, and amenities of the outpatient department, the course of treatment may vary from hospital to hospital. Therefore, it is advisable to read the hospital's reviews before visiting any OPD [8].

2.3 Relative

Family members (FMs) play a crucial role in patient care by aiding in decision-making, assisting the medical team with care, enhancing patient safety, providing support for in-home care, and meeting the needs of the patient's family or society at large.

2.4 Medical Doctor

Healthcare occupations include those in medical, dental, pharmacology, maternity, nursing, ophthalmology, audiology, psychology, occupational therapy, physiotherapy, athletic training, and other fields. This encompasses work done in public health, as well as primary care, secondary care, and tertiary care. Examples of healthcare providers include hospitals, physicians, nurses, clinics, healthcare facilities, medical experts, nutritionists, and dietitians, among others. Patients visit hospitals, clinics, and nursing homes for injuries or diagnoses.

2.5 Provider

A patient's request for therapy aimed at preventing, alleviating, treating, or curing a human disease, physical handicap, or damage is assisted by healthcare professionals. Healthcare providers connect patients with clinics, hospitals, and pharmacies to access medical services.

2.6 Manager

The health and medical service manager is in charge of planning and running hospital, clinics, and nursing homes, among other healthcare facilities. Instead of directly caring for patients, healthcare administrators make sure that hospitals run smoothly so that other healthcare workers can do their responsibilities.

2.7 Policymaker

A policymaker is someone who develops concepts and plans, particularly those that are carried out by businesses or the government. Policymakers include the mayor of a city, a school board, the board of directors of a firm, and or the prez. Plans that a government or organization follows are referred to as policy. The role of a policymaker is to make suggestions to other members of a government or stance body, such as non-governmental organizations, international organizations, or professional associations.

2.8 Supplier

A clinic is an economic entity since it offers medical services. It produces healthcare using production-related variables. The provision of health services is also examined using general microeconomic principles of production and cost. The quantity of changeable inputs determines the outcome of health services in the immediate term since technology is constant (medical and non-medical personnel). The hospital aims to generate the highest average output at the lowest average expense. The output generated ultimately depends on the optimal mix of technological equipment and manpower, both of which are unpredictable production factors. The law of profit maximization, which states that marginal income equals marginal cost, forms the foundation on which each hospital bases its output determination.

2.9 Quality Manager

A healthcare worker who manages quality improvement initiatives strives to raise the standard of patient care by enhancing patient outcomes, patient safety, and nursing management. They also aim to improve policies and procedures. Responsibilities include reducing healthcare-associated infections (HAIs), enhancing policies and procedures throughout the healthcare system, and leading initiatives focused on operational effectiveness and sustainability. Healthcare organizations, such as hospitals, clinics, and other facilities, may hire quality improvement managers who collaborate with hospital employees. These managers work either in the field or in offices, overseeing projects, paperwork, and administrative responsibilities that support organizational objectives.

2.10 Accreditation Staff

By being accredited, staff members (including managers, procurement experts, TLs, and consultants) demonstrate that they have been found to possess the necessary skills and expertise. As a government procurement expert, providing guidance on purchasing is crucial. Healthcare organizations utilize accreditation as a self-evaluation and external peer observation process to analyze their performance against predetermined standards and implement strategies to continually improve the patient system. With limited ownership and accountability for all staff members, including medical and paramedical professionals, it encourages comprehensive professional development, knowledge, and abilities in a systematic manner. Lastly, accreditation provides insurance companies and other intermediaries with an impartial empanelment procedure.

3 Methodology

IBM created the statistical software package SPSS Statistics, which includes features for managing data, analytic tools, multimodal predictive analysis, business analytics, and police cases. SPA Inc., developed by IBM and acquired in 2009, is the most advanced iteration marketed under the name IBM SPSS Statistics. It is common to practice utilizing "IBM SPSS Statistics Sciences" (SPSS), a collection of software tools for modifying, analyzing, and displaying data. SPSS supports various formats. To enhance the software's capabilities for data entry, statistics, or reporting, many add-on modules can be acquired. The core program is called SPSS Base. The SPSS Advanced and SPSS Linear Regression add-on modules are, in our opinion, the most important for statistical analysis. Additionally, SPA Inc. offers free tools that integrate with SPSS. Versions of SPSS for Windows (98, 2000, ME, NT, and Windows Vista) are also available. Windows 2000 is compatible with SPSS version 11.0.1. We are confident that the SPSS instructions provided in each chapter will still be applicable to the described experiments, even though more recent versions of SPSS may be available by the time this book is issued [9].

4 Result and Discussion

The Cronbach's alpha reliability result is displayed in Table 1. The model's total Cronbach's alpha score is 0.590, which denotes a 0.33% dependability level. The above 373% Cronbach's alpha value model from the literature review can be taken into consideration for analysis.

Table 2 Reliability statistic individual parameter Cronbach's alpha reliability results in Inpatient 0.253, Outpatient 0.277, Relative 0.357, Medical doctor 0.28, Provider 0.38, Manager 0.129, Policymaker 0.162, Supplier 0.39, Quality manager 0.338, and Accreditation staff 0.434.

Table 3 Descriptive statistics values for analysis N, range, minimum, maximum, mean, standard deviation, variance, skewness, and kurtosis. Inpatient, Outpatient, Relative, Medical doctor, Provider, Manager, Policymaker, Supplier, Quality manager, and Accreditation staff are also used.

Table 1 Reliability statistics

Reliability Statistics		
Cronbach's alpha	Cronbach's alpha based on standardized items	N of items
0.333	0.373	10

Table 2 Reliability statistic individual

	Total statistics
Item	Cronbach's alpha if item deleted
Inpatient	0.253
Outpatient	0.277
Relative	0.357
Medical doctor	0.28
Provider	0.38
Manager	0.129
Policymaker	0.162
Supplier	0.39
Quality manager	0.338
Accreditation staff	0.434

Table 3 Descriptive statistics

Descriptive statistics													
	N	Range	Minimum	Maximum	Sum	Mean		Std. Deviation	Variance	Skewness		Kurtosis	
	Statistic	Statistic	Statistic	Statistic	Statistic	Statistic	Std. error	Statistic	Statistic	Statistic	Std. error	Statistic	Std. error
Inpatient	31	2	2	4	92	2.97	0.135	0.752	0.566	0.054	0.421	−1.164	0.821
Outpatient	31	4	1	5	97	3.13	0.211	1.176	1.383	−0.134	0.421	−0.663	0.821
Relative	31	4	1	5	101	3.26	0.207	1.154	1.331	−0.126	0.421	−0.670	0.821
Medical doctor	31	4	1	5	84	2.71	0.223	1.243	1.546	0.262	0.421	−0.774	0.821
Provider	31	4	1	5	83	2.68	0.243	1.351	1.826	0.203	0.421	−1.205	0.821
Manager	31	4	1	5	89	2.87	0.206	1.147	1.316	0.268	0.421	−0.647	0.821
Policymaker	31	4	1	5	98	3.16	0.218	1.214	1.473	0.030	0.421	−1.052	0.821
Supplier	31	4	1	5	91	2.94	0.241	1.340	1.796	−0.230	0.421	−1.195	0.821
Quality manager	31	4	1	5	94	3.03	0.234	1.303	1.699	0.226	0.421	−1.112	0.821
Accreditation staff	31	4	1	5	89	2.87	0.201	1.118	1.249	−0.036	0.421	−0.573	0.821
Valid N (listwise)	31												

Table 4 Frequency statistics

Statistics		Inpatient	Outpatient	Relative	Medical doctor	Provider	Education	Manager	Policymaker	Supplier	Quality manager	Accreditation staff
N	Valid	31	31	31	31	31	31	31	31	31	31	31
	Missing	0	0	0	0	0	0	0	0	0	0	0
Median		3.00	3.00	3.00	3.00	3.00		3.00	3.00	3.00	3.00	3.00
Mode		3	3	3	3	1		2	2	4	2	3
Percentiles	25	2.00	2.00	2.00	2.00	1.00		2.00	2.00	2.00	2.00	2.00
	50	3.00	3.00	3.00	3.00	3.00		3.00	3.00	3.00	3.00	3.00
	75	4.00	4.00	4.00	4.00	4.00		4.00	4.00	4.00	4.00	4.00

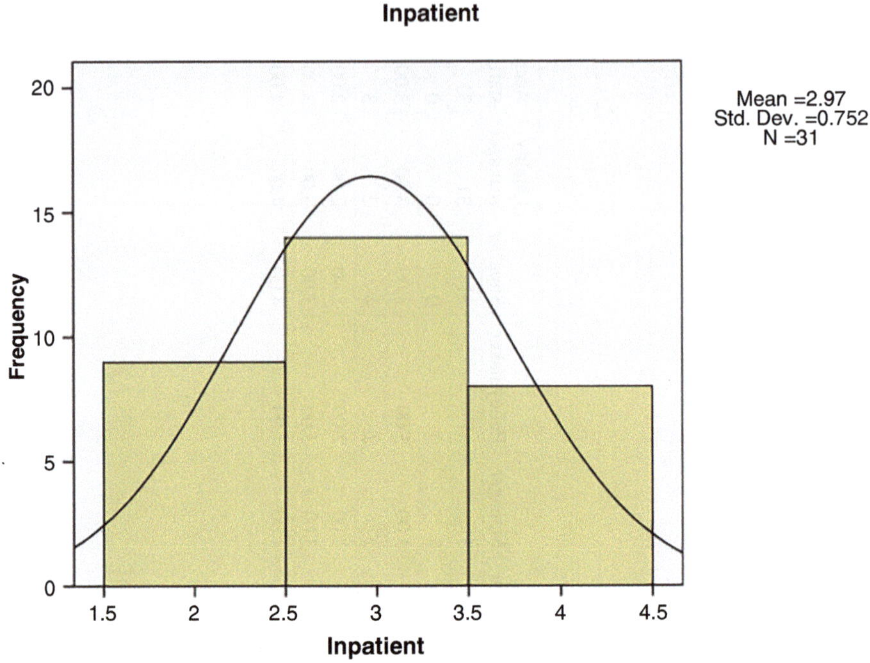

Fig. 1 Inpatient

Table 4 Frequency Statistics in Inpatient, Outpatient, Relative, Medical doctor, Provider, Manager, Policymaker, Supplier, Quality manager, and Accreditation staff values. Valid 31, Missing value 0, Median value 3.00, Mode value 3.

Figure 1 displays the histogram plot for the inpatient population; it is evident from the figure that the data are slightly right-skewed as more respondents selected the inpatient option between 2.5 and 3.5; however, aside from the value of 1, all other values are under the normal curve, indicating that the model is significantly consistent with a normal distribution.

Figure 2 displays the histogram plot for the outpatient population. It is evident from the figure that the data are slightly left-skewed because more respondents chose 3 for the outpatient. However, aside from the value of 2, all other values are under the normal curve, demonstrating that the model is significantly following a normal distribution.

Figure 3 displays the histogram plot for the relative, and it is obvious that the data are slightly left-skewed because more respondents chose the value 3 for the relative. However, aside from value 2, all other values are under the normal curve, demonstrating that the model is significantly consistent with a normal distribution.

Figure 4 displays the histogram plot for the medical doctor. It is evident from the figure that the data are slightly right-skewed as a result of more respondents selecting

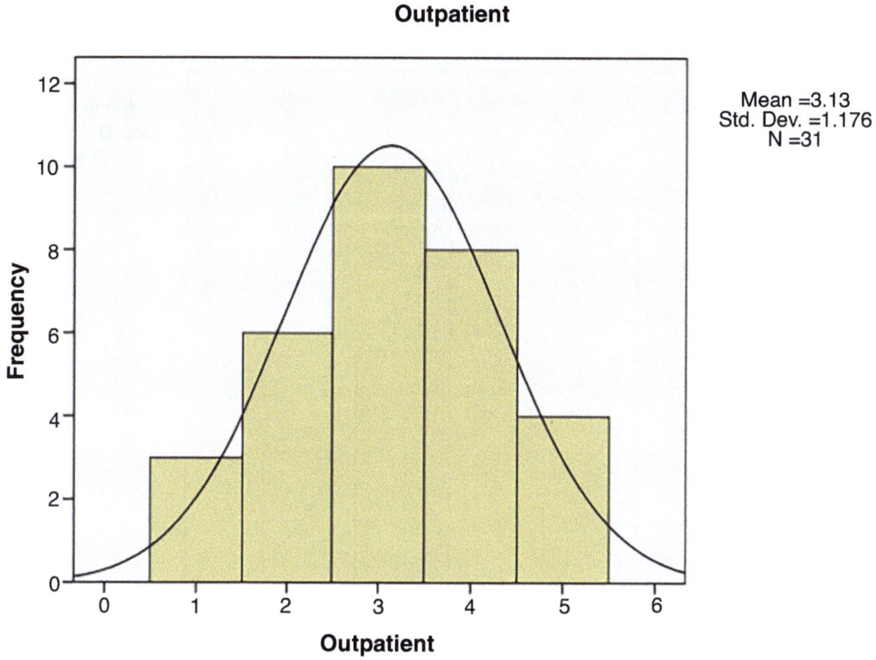

Fig. 2 Outpatient

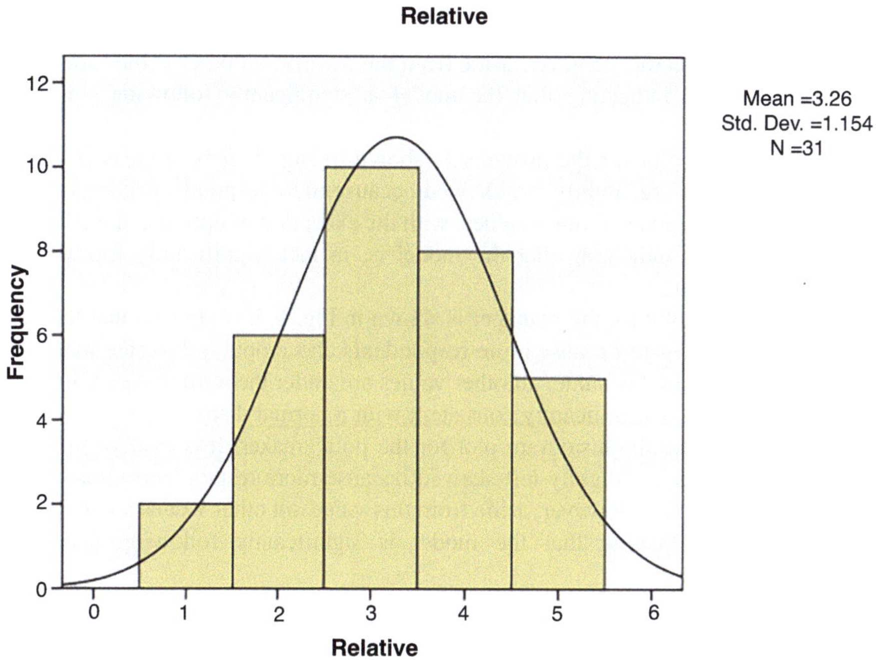

Fig. 3 Relative

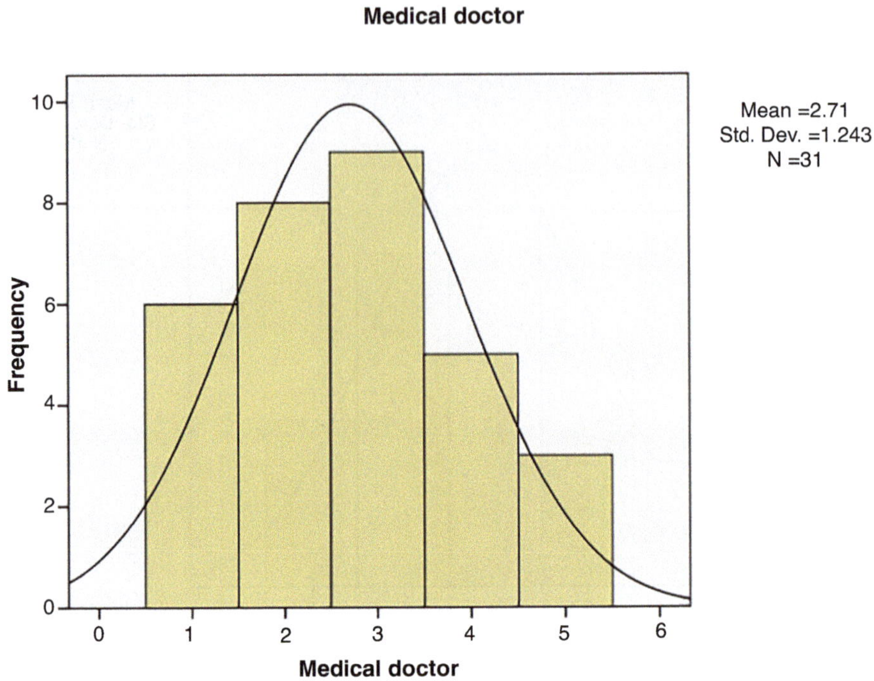

Fig. 4 Medical doctor

3 as the medical doctor; however, aside from the 3 value, all other values are under the normal curve, indicating that the model is significantly following a normal distribution.

The histogram plot for the provider is shown in Fig. 5. It is obvious from the figure that the data are slightly left-skewed because more respondents chose option 1 as the provider; however, other values, with the exception of option 2, are all under the normal curve, indicating that the model is, in fact, significantly following a normal distribution.

The histogram plot for the manager is shown in Fig. 6; it is obvious that the data are slightly left-skewed because more respondents chose option 2 for the manager; however, aside from this value, all other values are under the normal curve, indicating that the model is significantly consistent with a normal distribution.

Figure 7 displays the histogram plot for the policymaker. It is evident from the figure that the data are slightly left-skewed because more respondents chose value 2 for the policymaker. However, aside from this value, all other values are under the normal curve, indicating that the model is significantly following a normal distribution.

The histogram plot for the supplier is shown in Fig. 8; it is evident from the figure that the data are slightly right-skewed as a result of more respondents selecting 4 for

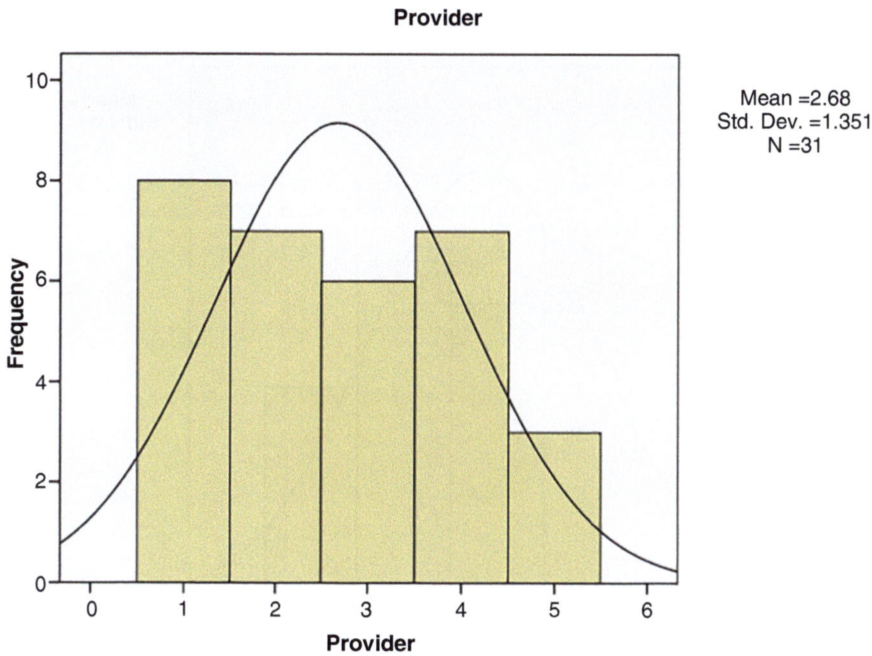

Fig. 5 Provider

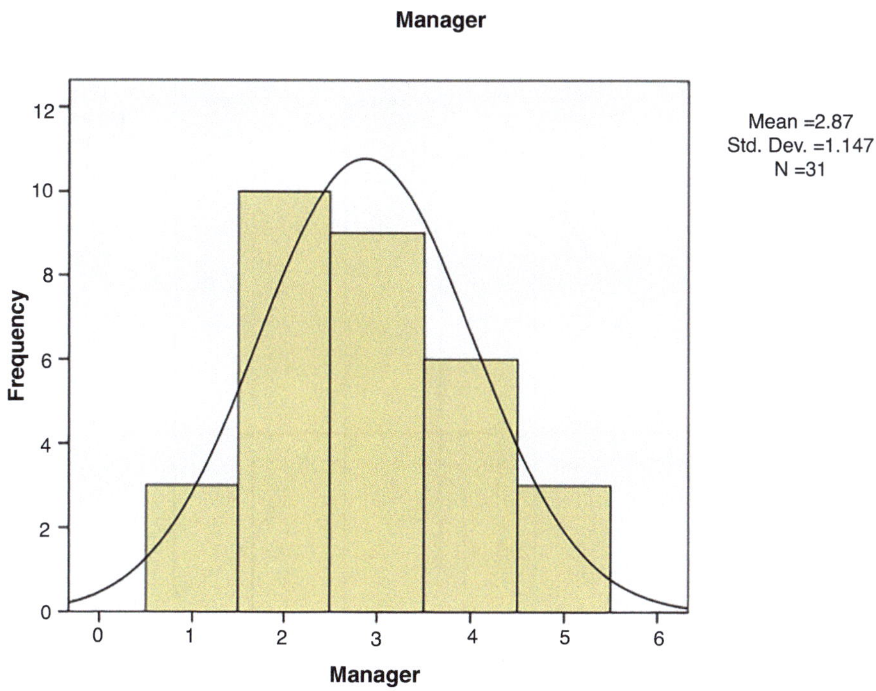

Fig. 6 Manager

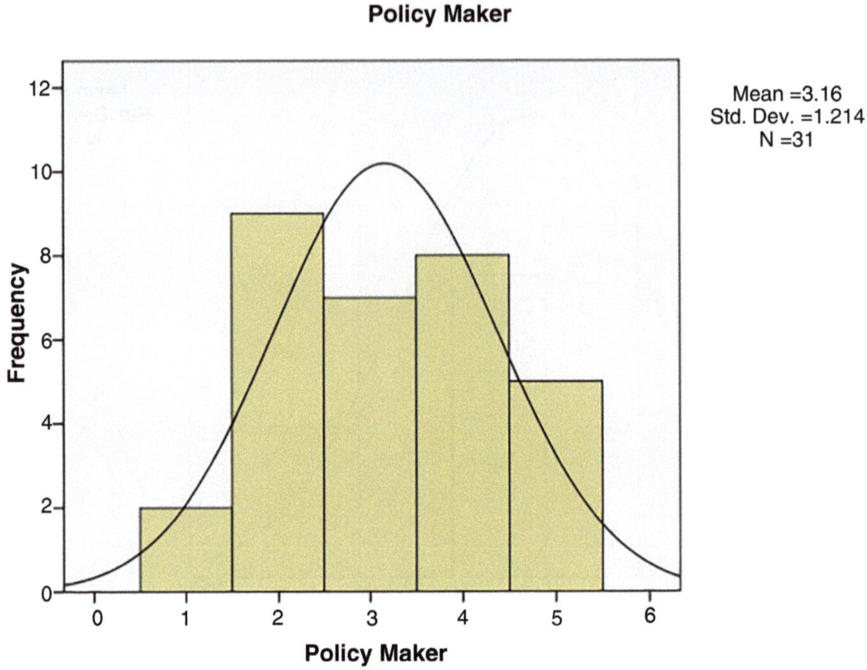

Fig. 7 Policymaker

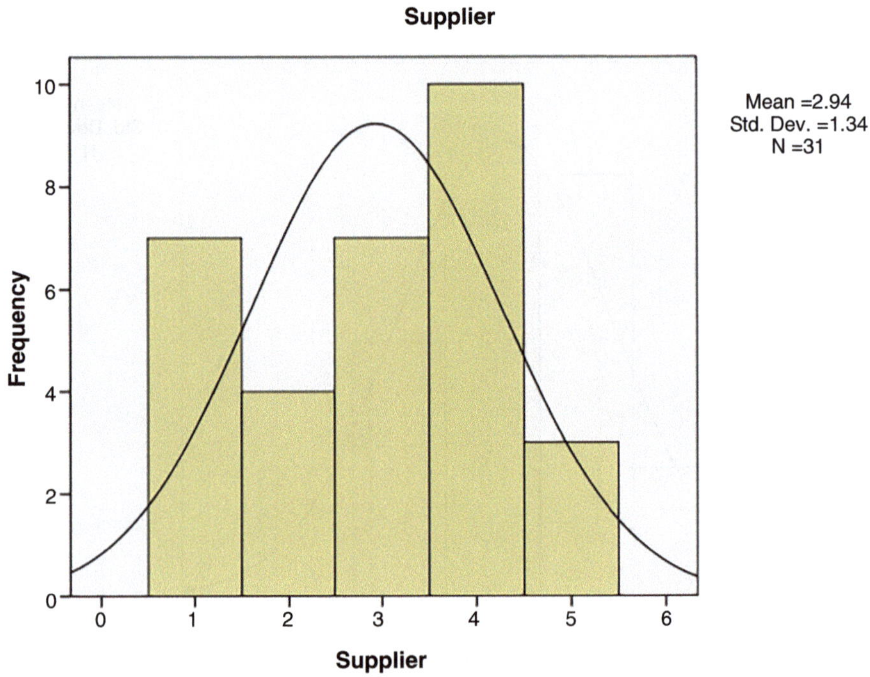

Fig. 8 Supplier

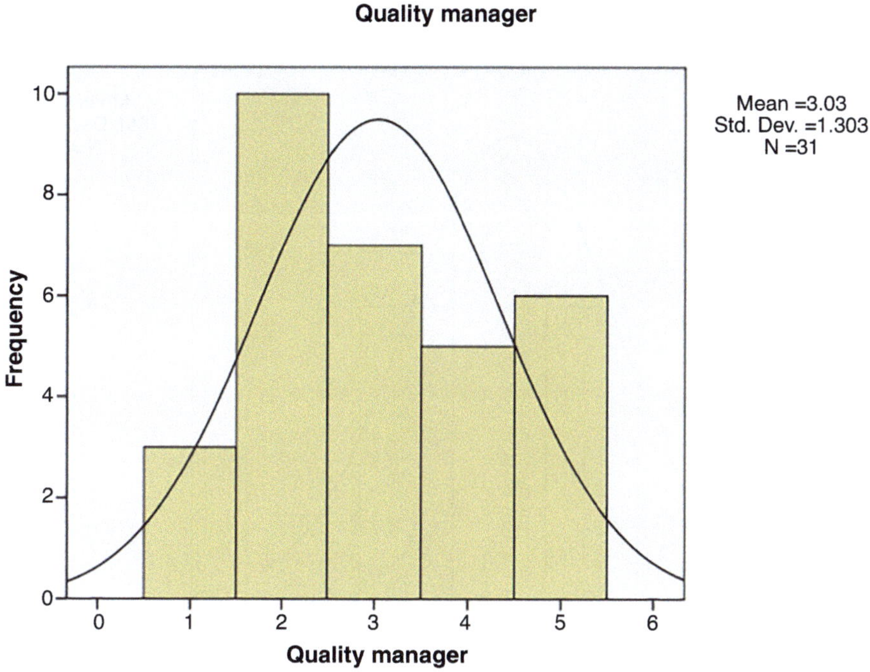

Fig. 9 Quality manager

the supplier; however, aside from the value of 2, all other values are under the normal curve, indicating that the model is significantly following a normal distribution.

Figure 9 displays the histogram plot for the quality manager. It is obvious from the figure that the data are slightly left-skewed because more respondents chose the value 2 for the quality manager. However, aside from the 2 value, all other values are under the normal curve, demonstrating that the model is significantly following a normal distribution.

Figure 10 displays the histogram plot for the accreditation staff. It is evident from the figure that the data are slightly right-skewed as a result of more respondents selecting 3 for the accreditation staff. However, aside from the 2 values, all other values are under the normal curve, demonstrating that the model is significantly consistent with a normal distribution.

Table 5 displays the relationship between the criteria of motivation for inpatient and outpatient care. Accreditation staff and quality managers have the lowest connection. Next, the association between the motivational factors for outpatient and inpatient care is strongest, while the correlation between accreditation staff is lowest. The next correlation is between the motivational factors for Relative them and a doctor, with Quality Manager having the lowest correlation. Next, Relative and Supplier have the strongest correlations for the motivation characteristics of medical doctors, whereas Supplier has the lowest correlations. Next, the Inpatient

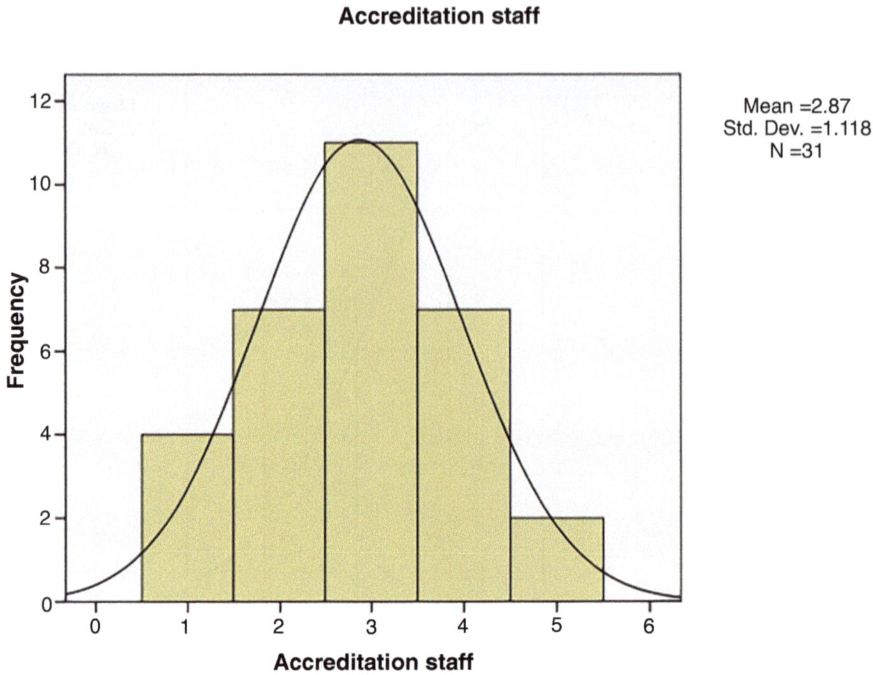

Fig. 10 Accreditation staff

and Provider motivation metrics have the highest correlations, whereas Supplier has the lowest correlations. Next, the relationship between the manager's and the policymaker's motivational parameters has the strongest connection, while the supplier's relationship has the lowest correlation. The second correlation is between the dimensions of motivation for Policymakers and Suppliers, with Accreditation personnel having the lowest correlation. Next, the Policymaker and Provider have the largest correlation and the lowest correlation between the incentive characteristics for the Supplier. Next, Quality Manager and Policymaker motivation factors have the strongest connection, whereas Relative motivation parameters have the lowest correlation. Next, using through correlation (Table 5) it was found that Diagnostics and Inpatient have the strongest correlation between the motivating metrics for Quality Managers, whereas Diagnostics has the lowest correlation and (Table 6 and Table 7) major influencing factors have been identified using the factor analysis and model fit also confirmed. (Tables 6 and 7).

Table 5 Correlations

Correlations	Inpatient	Outpatient	Relative	Medical doctor	Provider	Manager	Policymaker	Supplier	Quality manager	Accreditation staff
Inpatient	1	0.533**	−0.105	0.275	0.383*	0.381*	0.225	−0.134	−0.203	−0.203
Outpatient	0.533**	1	0	0.118	0.195	0.087	0.289	−0.121	−0.025	−0.241
Relative	−0.105	0	1	0.519**	−0.052	0.152	−0.007	−0.183	−0.227	−0.077
Medical doctor	0.275	0.118	0.519**	1	0.022	0.323	−0.1	−0.232	−0.117	−0.004
Provider	0.383*	0.195	−0.052	0.022	1	0.359*	−0.049	−0.362*	−0.145	−0.227
Manager	0.381*	0.087	0.152	0.323	0.359*	1	0.398*	0.125	0.003	−0.169
Policymaker	0.225	0.289	−0.007	−0.1	−0.049	0.398*	1	0.457**	0.313	−0.156
Supplier	−0.134	−0.121	−0.183	−0.232	−0.362*	0.125	0.457**	1	0.287	−0.028
Quality manager	−0.203	−0.025	−0.227	−0.117	−0.145	0.003	0.313	0.287	1	0.255
Accreditation staff	−0.203	−0.241	−0.077	−0.004	−0.227	−0.169	−0.156	−0.028	0.255	1

**Correlation is significant at the 0.01 level (2-tailed)
*Correlation is significant at the 0.05 level (2-tailed)

Table 6 Regression

Model summary

Model	R	R square	Adjusted R square	Std. error of the estimate	Change statistics				
					R square change	F change	df1	df2	Sig. F change
1	0.752[a]	0.565	0.379	0.593	0.565	3.030	9	21	0.017

ANOVA[b]

Model		Sum of squares	df	Mean square	F	Sig.
1	Regression	9.586	9	1.065	3.030	.017[a]
	Residual	7.382	21	0.352		
	Total	16.968	30			

[a]Predictors: (Constant), Accreditation staff, medical doctor, Policymaker, Provider, Outpatient, Quality manager, Relative, Supplier, Manager
[b]Dependent Variable: Inpatient

Table 7 Factor analysis

Total variance explained

Component	Initial Eigenvalues			Extraction sums of squared loadings		
	Total	% of variance	Cumulative %	Total	% of variance	Cumulative %
1	2.470	24.702	24.702	2.470	24.702	24.702
2	1.980	19.800	44.502	1.980	19.800	44.502
3	1.457	14.566	59.068	1.457	14.566	59.068
4	1.058	10.584	69.652	1.058	10.584	69.652
5	0.975	9.750	79.402			
6	0.706	7.063	86.465			
7	0.532	5.315	91.780			
8	0.339	3.387	95.167			
9	0.283	2.828	97.996			
10	0.200	2.004	100.000			

Extraction Method: principal component analysis

5 Conclusion

There are key aspects of high-quality medical care that are frequently observed. Every sample used in the current investigations has been reviewed. In these crucial dimensions, we discovered a few differences. SERVQUAL is most prevalent when focusing on dimensions in both developing and developed nations. It can be used as a foundational model to build upon or add more contextual dimensions. Observed factors contribute to these priorities regardless of the context. Resources can offer crucial information to help with decision-making on the quality of medical care. In other words, low-income patients may be as happy with the same level of service as patients with higher earnings if the healthcare is of high caliber. The overall satisfaction of patients is determined, not only using scores but also using means and gap analysis to directly measure patient satisfaction. There is much more to these satisfactions than meets the eye. These findings demonstrate the representativeness of the patients. Accordingly, the level of health service satisfaction at BEU HRAC indeed has a higher correlation score. Overall happiness is highly valued for its reliability. Outpatient satisfaction is a relative or determining dimension in the measurement. The study's findings demonstrate its dependability. The most decisive dimension is the dimension of hospital outpatients' satisfaction with Servqual aspects. However, the quality gap between perceptions and expectations of the believability component is larger, resulting in lower patient satisfaction. Reliability is a crucial dimension of health services. Patients place a high value on the reliability factor when determining the quality of health services. Additionally, they are satisfied with the worse reliability dimension on the SERVQUAL scales. High patient expectations could be to blame for this outcome. Managers and scientists should focus more on reliability to improve direct benefits and service quality, leading to indirect advantages for hospitals and patients. Stakeholders' opinions on what constitutes rising health services were solicited for this study. As standards for the quality of care provided by healthcare partners, the following seven quality rights in services were identified: proper treatment is done in the appropriate way to achieve the right results for the right person at the right time and the right price; your right guy must be present. The term "suitable services" refers to quality, patient-centered, appropriate, acceptable, required, accurate, safe, and effective health services. Effective use and proper servicing procedures are tied to doing things the right way. Accessible health facility options are provided if the space is adequate. Timely services are given when customers desire or need them. The ideal supplier demonstrates efficiency, responsibility, commitment, support, warmth, friendliness, and honesty in their hosting. Having the correct personnel ensures proper patient care. Fair pricing is reasonable for both the buyer and the supplier. Correct results suggest improved clinical outcomes. In other words, excellence is defined by sanitation. Meeting patient demands and satisfying providers require continuously meeting the patient's needs by offering effective, efficient, and cost-effective preventive care in accordance with shifting operational or customer demands. This definition considers the needs of the patient; therefore, quality should always be determined by shifting operational or customer demands. Quality in professional healthcare services is arbitrary and is a complicated, multifaceted concept. Its evaluation requires a generally accepted concept of health quality. The author discusses multivariate evaluation research that examined how stakeholders in healthcare perceived quality. The evaluation uncovered comprehensive hygiene and presented unimaginably uniform service quality. An evaluation strategy was implemented. Quality healthcare services have varied meanings for customers, providers, managers, and policymakers, as revealed by multivariate analysis. Managers and practitioners should place a high emphasis on the aspects of health quality that are significant to each group involved in service delivery. They must succeed on these levels if they hope to satisfy their clients. This study illustrates the intricacy of health quality and advances our knowledge of it.

References

1. Haron, S. N., Hamida, M. Y., & Talib, A. (2012). Towards healthcare service quality: An understanding of the usability concept in healthcare design. *Procedia-Social and Behavioral Sciences, 42*, 63–73.
2. Endeshaw, B. (2022). Healthcare service quality-measurement models: A review. *Journal of Health Research, 35*(2), 106–117.
3. Lee, P., Khong, P., & Ghista, D. N. (2006). Impact of deficient healthcare service quality. *The TQM Magazine, 18*(6), 563–571.
4. Fatima, I., Humayun, A., Iqbal, U., & Shafiq, M. (2019). Dimensions of service quality in healthcare: A systematic review of literature. *International Journal for Quality in Health Care, 31*, 11–291.
5. Fatima, T., Malik, S. A., & Shabbir, A. (2018). Hospital healthcare service quality, patient satisfaction and loyalty: An investigation in context of private healthcare systems. *International Journal of Quality & Reliability Management, 35*(6), 1195–1214.
6. James, T. L., Villacis Calderon, E. D., & Cook., D. F. (2017). Exploring patient perceptions of healthcare service quality through analysis of unstructured feedback. *Expert Systems with Applications, 71*, 479–492.
7. Al-Damen, R. (2017). Health care service quality and its impact on patient satisfaction "case of Al-Bashir hospital".
8. Lee, D. H., & Kim, K. K. (2017). Assessing healthcare service quality: A comparative study of patient treatment types. *International Journal of Quality Innovation, 3*(1), 1–15.
9. Chahal, H., & Kumari, N. (2010). Development of multidimensional scale for healthcare service quality (HCSQ) in Indian context. *Journal of Indian Business Research, 2*(4), 230–255.

Value-Added Products from Agricultural Waste

Tinto Tom and E. Eswara Reddy

Abstract The possibility of utilizing agricultural waste to create value is covered in this article. Agricultural waste is a significant environmental problem that causes soil, water, and air pollution. Yet this waste can be used to create various products with additional value, such as biofuels, bioplastics, and animal feed. The numerous technologies and techniques used to transform agricultural waste into value-added goods are also explored in the study, along with the various forms of agricultural waste and their composition. The study also emphasizes the advantages of agricultural waste utilization from an economic and environmental standpoint and any difficulties and possibilities that may arise from this strategy. Ultimately, this chapter highlights the significance of agricultural waste value addition for circular economies and sustainable agriculture.

Keywords Agricultural waste · Pollution · Value addition

1 Introduction

India is one of the world's most populated countries, with the majority of the population working in agriculture. The Indian agriculture sector has a great history in its development. Agriculture started in India with the Indus Valley Civilization. Rice and cotton were the major crops cultivated during that period. Later, animal husbandry also developed as part of agriculture. At present, agriculture is the backbone of the Indian economy, and it is considered a prerequisite for economic development. It is the foundation of the entire structure of the economy. If we connect the economy in terms of the galaxy, agriculture is the center, and all other sectors revolve around agriculture. Without the agriculture sector, the development of economic growth is night dream.

T. Tom (✉) · E. E. Reddy
CMR University, Bengaluru, Karnataka, India
e-mail: tinto.tom@cmr.edu.in

J. M. K. P. et al. (eds.), *Business Resilience and Digital Technology in the Post-Pandemic Era*, Contributions to Environmental Sciences & Innovative Business Technology, https://doi.org/10.1007/978-3-031-48075-1_7

The main problem of an economy is of economizing scarce resources. Scarcity takes place due to unlimited wants and limited resources. It is a fact that all economies face this issue. So the economies must use their resources appropriately to fulfill their unlimited wants. All human activities lead to waste generation. Wastes are unwanted or unusable materials. As per Wikipedia, waste is any substance discarded after its primary use or is worthless, defective, and without use. Even after agricultural production, a lot of waste is generated that we may dispose of in different ways. Since our resources are scarce, we should utilize the material even after its primary use, or worthless or defective thing, because these wastes may create value and be useful to satisfy our needs.

In the Indian scenario, more than 50% of the total population works in the agriculture sector, but the agriculture sector's contribution toward GDP is less than 15%. There are many reasons for the poor performance of the agriculture sector. But still, a lot of ways are there to develop the agriculture sector. One such way is to give priority to agricultural waste. This agricultural waste management and reuse may make many changes in the agriculture sector itself. Some methods are prevailing in society, like biogas plantation and vermiculture, etc. But still, people are not adopting and practicing these methods in their fields. It may be because of the lack of awareness or support from the authority. Proper utilization of agricultural wastes can make many changes in the economy concerning the contribution toward GDP, and natural fertilizer has a lot of scope and demand in the economy. It can generate more employment opportunities and help make pesticide-free crops that are good for human health. It may become a source of capital formation, and many more positive outcomes are there from agriculture waste management.

2 Literature Survey

Hsu [1] studied the cost-benefit analysis for recycling the agricultural waste in Taiwan. The author conveyed that agricultural wastes can be recycled differently, but the major concern is cost-effectiveness. People, companies, or governments will consider a profitable way to recycle agricultural waste. There are three major agricultural waste recycling methods: biogas power generation, compost, and biomass fuel. This study result conveys that these methods are profitable. Higher cost-effective methods include composting chicken manure along with other agricultural wastes and producing biomass fuel rods from rice straw. To encourage the recycling of agricultural waste, policy support and tariff adjustments are crucial. We can use these techniques if they are effective in Taiwan and India. It could be a solution to a variety of issues that are prevalent in India.

Singh et al. [2] studied strategies and prospects to integrate sugarcane production technologies for enhanced cane and sugar productivity targeting to increase farmers' income. The study conveys that the government of India plans to increase the income of agriculturists. It can be achieved by improving cost-effective technologies, educating farmers, bringing the laboratory to land, etc. The targeted group here is sugarcane farmers. Developing technologies to convert wastes into resources may help them gain additional income.

Sethuraman and Kavitha [3] studied about vermicomposting of green waste using the earthworm *Lumbricus rubellus*. As per the study, India produces huge amounts of organic waste yearly. These wastes come from different sources such as agriculture, urban, industries, and domestic activities. Management of these wastes is essential for economic and environmental reasons. The study conveys that earthworms can convert these biodegradable green wastes into value-added compost. It will improve soil fertility, and plants and vegetables will grow without chemical fertilizers. It has a lot of demand in the present situation.

Mandavgane and Kulkarni [4] conducted a study on the valorization of cow urine and dung. The study indicates that cow is a common livestock of Indian farmers. It contributes around 16% of their total earnings. Farmers in India follow a hybrid farming system, including crop and livestock cultivation. Studies convey that milk, cow dung, and urine are valuable resources and can be converted into value-added products. Even though it can be used as manure, the pesticide will also rejuvenate soil fertility. These prepare medical and cosmetic items.

Chilakamarry et al. [5] studied opportunities and challenges in solid-state fermentation for the bioconversion of agricultural wastes to value-added products. The study projects that the high volume of solid waste becomes a common problem everywhere. To achieve sustainable development, we should adopt a green approach. Agricultural wastes can be converted into profitable products through solid waste fermentation. Agricultural waste has high organic content that is used as potential substrates to produce value-added products through solid waste fermentation.

Babla et al. [6] researched on value-added products as soil conditioners for sustainable agriculture. One of the frequent issues resulting from the extensive use of fertilizers is soil degradation. It resulted in the loss of soil fertility and organic matter. Waste production was hastened by the intensification of agriculture, urbanization, and industry. Global developments to better utilize waste materials to create value-added products, including soil conditioners, are being driven by sustainable food production. For the health of the soil and the growth of plants, soil conditioners are crucial. We will be able to use fewer chemical fertilizers because of it.

Donner et al. [7] researched on a new circular business model typology for creating value from agro-waste. According to the study, in order to transition from a linear to a circular economy, we need novel business models, such as reverse logistics, fresh perspectives on customer-supplier interactions, and new organizational and marketing tactics at the intersection of diverse value chains. It gives managers choices for positioning and modifying their corporate strategies. It emphasizes how biomass has the potential to be used for greater added-value products before being used as an energy source.

Sadh et al. [8] reviewed agro-industrial wastes and their use of solid-state fermentation. In accordance with the study, agricultural leftovers are full of bioactive substances. The manufacture of goods like biogas, biofuel, mushrooms, etc. can be done using this agricultural waste as a substitute source. By using agricultural waste as a raw material, production costs and pollution levels are decreased. Through solid-state fermentation, even this agricultural waste can produce biofuels, enzymes, vitamins, antioxidants, animal feed, antibiotics, and other compounds.

2.1 Value-Added Products from Agricultural Waste

Agricultural waste management has a lot of scope in the present scenario. It may become a solution for most of the problems in our system. Many people know the income or value generation possibilities from agriculture waste management. But they are not ready to take the initiative or risk toward it. All resources, including agricultural waste, should be utilized to their maximum potential. We should not waste things because the resources are scarce. If we are not utilizing it properly in the later stage, there may be shortages of those valuable items. Some examples of value generation from agricultural wastes are discussed below.

2.2 Use as Fertilizers

Directly utilizing animal manure as fertilizer has a direct impact on reducing the required input energy on the farm. In fact, manure can contribute significantly to the availability of essential nutrients, with percentages of 19, 38, and 61% for phosphorus, nitrogen, and potassium, respectively, in comparison to chemical fertilizers. However, the use of confinement manures on a large scale for fertilization comes with notable energy expenses related to transportation, distribution, storage requirements, and concerns related to odors and potential groundwater pollution. Notably, poultry manure is rich in phosphorus, which benefits crop growth and overall productivity. Its effectiveness is even greater when employed alongside mineral phosphorus fertilizers in agricultural practices. Beyond nutrient enrichment, manure enhances soil quality by improving its physical attributes, water retention capabilities, and structural stability. This, in turn, elevates soil fertility by augmenting its capacity to retain nutrients (or cation exchange capacity).

2.3 The Anaerobic Process

Agricultural residues, particularly manures, offer a valuable resource for generating methane gas. This gas serves effectively in heating applications such as broiler operation, water heating, and grain drying. The anaerobic digestion of agricultural waste involves a two-step microbial fermentation process to produce methane-rich gas. Acid-forming bacteria initially break down volatile solids into organic acids, which are then utilized by methanogenic organisms to yield methane-rich gas. On average, the produced gas consists of 50–70% methane, 25–45% CO_2, 0.5–3% N_2, 1–10% H_2, and trace amounts of H_2S. Its heating value ranges from 18 to 25 MJ/m3. Although the anaerobic digestion system has initial high costs and concerns over methane's explosiveness, its benefits far outweigh these limitations. It effectively manages and deodorizes large quantities of poultry, swine, and dairy waste while preserving some fertilizer value in the residual waste.

2.4 Adsorbents for Heavy Metal Removal

Addressing the issue of heavy metal contamination resulting from industrialization and urbanization is a global challenge. Unlike organic pollutants, heavy metal ions (such as copper, cadmium, mercury, zinc, chromium, and lead) do not degrade into harmless substances and pose threats to various forms of life. Adsorption presents a viable method for removing heavy metal ions from wastewater. Numerous studies have explored activated carbon for treating heavy metal-laden wastewater. Agricultural residues have emerged as an economical alternative for adsorptive treatment of effluents containing heavy metals.

2.5 Animal Feed

In many developing nations, a major challenge in animal feed is the limited availability of protein sources. Despite extensive efforts to find alternative supplements, this scarcity persists. Agricultural residues, while rich in fiber, lack sufficient protein, carbohydrates, and fats. This poses a problem for the traditional method of boosting livestock production by adding grains and protein concentrates to grass and pasture. This approach might not adequately meet the growing demand for meat protein in the future. Moreover, as protein and grains are also needed for human consumption, there's a competition for these resources. Utilizing food leftovers to feed farmed animals presents a promising solution to address these complex issues.

2.6 Handicrafts

Red mud is traditionally used to produce mud cups or kulhad. The supply of this mud has been impacted by ongoing source depletion. Hence, it is necessary to discover alternatives to substitute red soil entirely or in part while creating kulhad. To turn waste into a product with additional value, corn cobs, which would otherwise be a waste product, are used to make mud cups. Mud and agricultural biomass leftovers can be used in part to make cups. Using biomass in this way offers an alternative method of removing leftover corn cobs.

2.7 Silage of Pineapple Fruit Leftovers as a Source of Animal Feed

Almost 1.3 million tons of non-edible pineapple fruit residue are available in India each year and are discarded. The pineapple fruit residue (PFR) is prone to fungal growth and spoiling within 2 days because of its high moisture content (65–70%) and total sugar content (>50%). Through silage technology, the quality of PFR was improved, allowing for the conversion of this fruit leftover into animal feed. The PFR was divided into pieces measuring 1–2 inches long and compressed in drums or bags with a moisture level of roughly 65% using a 4:1 (w/w) ratio of pineapple leafy crown and fruit peels. Good PFR hay was made over 20 days, and its nutritional content was superior to green maize fodder. Sheep were fed a total mixed ration of PFR silage that had no negative effects and supported the necessary growth. A study with dairy cows fed a diet based on PFR silage revealed an improvement in daily milk production of nearly 20% and a 0.6 unit increase in fat content.

2.8 Biscuits, Bread, and Rusk: Fortified Baked Goods from Cabbage Waste

Field waste from cabbage leaf production in India amounts to over 4.0 million tons. Large amounts of healthy nutrients, such as 390 mg of vitamin C per 100 g of food, 2.63% protein, and 6537 g of total carotenoids per 100 g, are present in these wastes. The potential for using these discarded leaves to fortify food is enormous. A procedure has been standardized for everyday foods with more nutritional content than usual, such as bread, biscuits, and rusk.

2.9 Banana Central Core Stem Pickle and Juice

The stem from banana plants is a crop waste with little commercial value. After the banana crop is harvested, around 25–30 tons of pseudo-stem debris are produced per hectare, of which roughly 5–7 tons of the central core can be removed. Poovan and Karpuravalli varieties work well with center stems. It is transformed into a high-value product and a central core stem pickle.

For most culinary preparations, the core stem component is employed. Manually or mechanically, the juice is removed from the central core stem. The stem is high in potassium and fiber. On average, 80–95% of the juice from the center stem of commercial banana cultivars could be recovered.

2.10 Paper Plates Made from Natural Biomass Fibers

Banana pseudo-stem fibers and sugarcane bagasse can be used to create disposable paper plates, thanks to a technology created by ICAR-CIRCOT. A revolutionary bee wax or sugarcane wax-based coating method replaces the traditional plastic coating on paper plates. The created paper plates had a decent aesthetic appearance and enough strength to support the food items. Banana farmers, companies that harvest banana fiber, sugarcane businesses, incubators, and start-ups are some potential interest holders.

2.11 Production of Biochar from Coconut Wastes

In India, the biomass wastes produced by coconut plantations and small-scale industries based on coconuts are nearly 25 million MT yearly. Due to their high lignin content, they are an excellent feedstock for biochar production. A solid, carbon-rich product with an added value known as biochar is produced by pyrolyzing waste materials from livestock, forestry, and other sources in low- or no-oxygen settings. This led to the development of a method at ICAR-CPCRI that uses a straightforward charring kiln to pyrolyze coconut wastes, such as tender coconut husks, adult coconut husks, coconut petioles, etc. into biochar. The coconut wastes can be converted into biochar with a weight-to-weight turnover of 40–50%.

2.12 Chitin and Chitosan from Prawn Shell Waste

Chitin is the second most prevalent biopolymer, a key structural element of arthropods' exoskeletons. Commercial sources for this commodity include prawn shells. Chitin, chitosan, glucosamine, and numerous more derivatives are produced from chitin. Several industries use chitin and its related products, including healthcare, manufacturing, agriculture, biotechnology, and nutraceuticals. More than 1.25 lakh tons of head and shell debris are produced annually by the shrimp-processing industry in India. Shell and head waste from shellfish processing is a hazard to the environment. Over 50% of the raw material is wasted during processing. Now, discarded shrimp shell wastes have the potential to yield 7000 tons of chitin. ICAR-CIFT has refined technologies for various uses in the food and medical industries.

2.13 *Fish and Shellfish Head Waste Is Used to Make Protein-Based, Bioactive Edible Films*

Due to the country's massive annual processing of shrimp for export, shrimp head trash is quite prevalent in India. On a wet weight basis, shrimp head waste has a protein content above 12%. Fish head, which includes proteins, lipids, and other minor nutrients, is also frequently discarded during processing, accounting for nearly 35–40% (w/w) of solid waste. Due to structural complexity and increased enzyme breakdown, getting proteins from fish and shrimp heads is extremely difficult (in shrimp heads). ICAR-CIFT developed the pH and enzyme-assisted extraction method to recover the protein in the form of physiologically active peptides. These protein derivatives have been discovered to benefit the creation of bioactive edible coatings or films.

3 Conclusion

It is possible to add value to agricultural waste, which was once thought of as a waste product, through valorization. A more sustainable agricultural business, decreased environmental damage, and economic gains can all result from using agricultural waste in value-added products. Farmers, businesses, and researchers have new options to investigate, thanks to the continually developing technology and techniques for turning agricultural waste into value-added goods. The scalability of technologies, the efficiency of the processes, and the possibility of competition with food production still need to be resolved. Overall, the value-adding of agricultural waste is a significant step toward a more circular and sustainable economy. More research and funding are required in this area to realize its full potential.

References

1. Hsu, E. (2021). Cost-benefit analysis for recycling of agricultural wastes in Taiwan. *Waste Management, 120*, 424–432. https://doi.org/10.1016/J.WASMAN.2020.09.051
2. Singh, P., Singh, S. N., Tiwari, A. K., Pathak, S. K., Singh, A. K., Srivastava, S., & Mohan, N. (2019). Integration of sugarcane production technologies for enhanced cane and sugar productivity targeting to increase farmers' income: Strategies and prospects. In *3 Biotech* (Vol. 9(2)). https://doi.org/10.1007/s13205-019-1568-0
3. Sethuraman, T. R., & Kavitha, K. V. (2013). Vermicomposting of green waste using earthworm Lumbricus rubellus. *Nature Environment and Pollution Technology, 12*(2), 371.
4. Mandavgane, S. A., & Kulkarni, B. D. (2020). Valorization of cow urine and dung: A model biorefinery. In *Waste and biomass valorization* (Vol. 11(3)). https://doi.org/10.1007/s12649-018-0406-7

5. Chilakamarry, C. R., Mimi Sakinah, A. M., Zularisam, A. W., Sirohi, R., Khilji, I. A., Ahmad, N., & Pandey, A. (2022). Advances in solid-state fermentation for bioconversion of agricultural wastes to value-added products: Opportunities and challenges. In Bioresource Technology (Vol. 343). doi:https://doi.org/10.1016/j.biortech.2021.126065.
6. Babla, M., Katwal, U., Yong, M. T., Jahandari, S., Rahme, M., Chen, Z. H., & Tao, Z. (2022). Value-added products as soil conditioners for sustainable agriculture. In *Resources, conservation and recycling* (Vol. 178). https://doi.org/10.1016/j.resconrec.2021.106079
7. Donner, M., Gohier, R., & de Vries, H. (2020). A new circular business model typology for creating value from agro-waste. *Science of the Total Environment, 716*, 137065. https://doi.org/10.1016/j.scitotenv.2020.137065
8. Sadh, P. K., Duhan, S., & Duhan, J. S. (2018). Agro-industrial wastes and their utilization using solid state fermentation: A review. In *Bioresources and bioprocessing* (Vol. 5(1)). https://doi.org/10.1186/s40643-017-0187-z

Brand Revitalization: An Outlook

Shanmugha Priya Pon

Abstract Changes in the marketplace can motivate brands to quickly change their reputations, and companies have to endlessly appraise their brands from customer-centric points of view. Refreshing a brand name can revive the worth of a brand and spur on innovation improvements. Constructing an impressive brand takes time, perseverance, and ongoing efforts, where the benefits are some of the most important parts of the instantaneous recognition of a business in the customer's mind, a reputation or competency and superior quality that the belief that the brand's promises are genuine and not just slogans, etc. When a brand fails to find a market, its parent company must choose whether to keep the brand as is or rejuvenate it by removing its products and/or services from consumers' minds. Breathing new life into a failing brand is frequently more effective than rejuvenating, it because the current brand has intrinsic value to the company. If a legacy brand is to stay on top nowadays, it must remain meaningful to customers. Revitalizing an established brand requires reassessing how representative its offerings are to its target market.

Keywords Brand revitalization · Brand audit · Repositioning

1 Introduction

Brand revitalization is the wide-ranging renovation of a brand, including its position and its unique brand name. Refreshing a lagging brand can be tremendously rewards. Over time, brands develop unsavory reputations among customers. Some people even express themselves through brands that remind them of recognizable events or places, and older brands represent a level of dependability that innovative brands cannot match. Recognized companies time and again turn to promotion to rejuvenate their individuality and enhance their significance in their marketplace. If

S. P. Pon (✉)
Department Commerce (Foreign Trade), PSG College of Arts and Science, Coimbatore, India

J. M. K. P. et al. (eds.), *Business Resilience and Digital Technology in the Post-Pandemic Era*, Contributions to Environmental Sciences & Innovative Business Technology, https://doi.org/10.1007/978-3-031-48075-1_8

the perception or quality of a company's goods/services falls over time, it may be time to reconsider the brand.

2 Brand Revitalization: An Outlook

From time to time, a business needs to reinvent a diminishing brand or an assortment of such brands. Shifts in the promotion atmosphere, opponents' strategies, consumer behaviors, the company culture, and other factors are capable of corroding a brand long term. Brand revitalization is a strategy aimed at building up a brand's strengths and to customize or recite and to customize or recategorize its brand portfolio. Revitalizing a brand allows consumers to experience anew the brand's offering, exposing them to new products in new places. Brand revitalization takes several forms:

- Awareness expansion: Increases customers' exposures to a brand's image.
- Brand enhancement: Increases the brand's potency, distinctiveness, and attractiveness, aiming to hold onto its current customers or win back lost customers.
- Innovation equilibrium and older markets requires manifold marketing programs, keeping the brand current, and generating new outlets to increase accessibility.

3 Revitalization and Its Implications

In order to refresh a brand, or recoup brand value, the brand must be stripped down to its unique sources of value. The aim here is to precisely identify which parts remain attractive associations for customers when repositioning the brand. Brand audits are carried out over and over again to identify these associations. A brand audit is a wide-ranging assessment of the fitness of a brand to determine its sources of brand equity from both a business and a consumer perspective. This gives the company indicators that inform long-standing strategic decisions and decisions about maintaining or changing something to improve the position or future direction of the brand. Properly positioning a brand requires considering the brand's attractiveness and its ability to deliver an assortment of achievable associations. According to the brand equity structure, two universal approaches are suitable to increase the depth and/or breadth of brand awareness via ornamental brand recognition and appreciation, and increase attractive qualities such as the strength expediency, and distinctiveness of the brand's image.

4 Intensifying Brand Awareness

Even a vanishing brand is usually still recognized by customers within a certain amount of time, which means that there is still time to increase the brand's awareness. The breadth of brand awareness can be increased to ensure that consumers do not overlook it and that they then consider buying from that brand when it meets their desires and requirements. A brand aiming to have adequate receptiveness and a positive brand image can be reinvigorated by finding and promoting groundbreaking sources of brand-name awareness. Sources of brand awareness can be created by employing tactics that increase exposure. Exposure can be increased by increasing consumption frequency (how frequently the brand is used).

5 Identifying Innovation and Ways to Employ the Brand

When revitalizing a brand, finding the product innovation is crucial as it tends to change the overall blueprint of the product. Managing brands over a longer period of time is about strengthening or, if necessary, revitalizing brands. Strengthening brands ensures improvements in product design, automation, and merchandizing and places particular emphasis on user and usage images. The entire product range is therefore subject to new, innovation-intensive adjustments that force customers to remember the established brands and at the same time be open to using the new offerings.

6 Identifying Additional or New Usage Opportunities

There are many customs that increase the likelihood of use. To first identify additional or new opportunities for consumers to use the brand beyond remaining in the same primary system, a marketing campaign must include the following: inform customers of the brand's suitability and location more frequently in accessible or novel situations; and encourage customers to actually use the spot in individual situations. This increases the frequency of use of resources with the aim of changing the way customers perceive their use, starting with the realistic use of these resources. One tactic to speed product substitution—which is particularly necessary for products manufactured in response to market demands—is to tie product substitution to a specific occasion, holiday, or time. There is another way to give customers more information about when the product was first used (or should be replaced) or what its current performance level is. If actual usage of a product falls short of ideal or recommended usage, there is another method to increase usage. Under these conditions, consumers must be convinced of the benefits of using the product more frequently and any barriers to this use must be removed, says Brand Revitalization:

An Outlook. Product packaging and design can improve the usability and practicality of a product.

7 Improving the Brand's Image

A original marketing strategy may be needed to improve the strength, popularity, and uniqueness of the brand associations that make up the image. When getting something in the right position or recommitting to accessible positioning, faded positive relations be able to become known, which means over the time the brand may get out of the market and sometimes it may vanish from the minds of the customer and from the market. To increase the potency, ubiquity, and distinctiveness of the brand associations that comprise the image, a novel marketing approach might be needed. Relocating or focusing on more approachable positioning may allow for the recognition of fading positive relationships, which will eventually cause the brand to fade from the market and occasionally vanish from consumers' perceptions. Furthermore, in order to achieve the stated goal, some unproductive relationships must be eliminated, and additional positive relationships may need to be formed as needed. To make the brand stand out more, brand revival might need the development of additional indisputable differentiators. In order to achieve equivalency across various input image proportions, it might also be required to move the mark. Rejuvenating established brands involves updating them with more contemporary features, consumer profiles, or suitable procedures. A brand refresh can include creative merchandise, fresh packaging, promotions, and advertising. Determining the target market and the competitive landscape is necessary for making positioning decisions. The target market for a brand usually does not include every possible segment that could make up the market as a whole. These excellent products are available under several company brands.

8 Harmonizing New and Old Markets

To increase sales, companies must be capable of deciding to target numerous marketplace segments. The improving the existing brand should increase the sales and must build a new market in such a way that it should capture all segments of the market and customers too, which ultimately bridges or harmonizes between new and old brands. There are three approaches to attracting new customers:

- Manifold marketing interactions plan—single loom toward drawing new market division to a brand at the same time as fulfilling existing divisions is to generate divide promotion drive and interactions series intended for all division.

- Expansion of the brand name along with sub-brands: in order to attract more creative consumers and keep the brand modern and relevant, the company must introduce a new sub-brand or bring in a mixed bag of extensions.
- New distribution points—in order to enter a new market segment, the product may need to be made more available to that group.

9 Guidelines to Be Followed During Brand Revitalization

Stimulating an existing brand means civilizing or altering the existing observation. The quantity of alteration depends on the brand value and the goals of the customer. The following are important guidelines to follow at the time of revitalization:

- Identifying the significant changes in the market and in competitors' activities.
- Identifying the significant changes in customer perceptions and expectations.
- Redefining the brand vision and goals into something that is understood and supported.
- Redefining corporate and brand values: brand realization.
- Creation of a new corporate identity that faithfully reflects the brand direction.
- Globalization has made branding increasingly important in creating competitive advantages for businesses.

Over the last decade, more and more companies have made acquisitions of established brands as an alternative to building new brands.

10 Reasons for Revitalization

Rebranding and brand revitalization are not mundane tasks to be undertaken without due consideration and are supported by legitimate influence. Some of the extended reasons for rebranding consider the following explanatory reasons to force the company to revitalize its brand: transformation, once a brand turns out to be old-fashioned and is in danger of slowly disappearing from the market; distinction through well-constructed antagonism, otherwise rapidly changing environment; overcrowding or lack of management relative to competitors; dealing with greater price competitiveness; dealing with the demands of brand globalization; dealing with a brand mergers or acquisitions; improving competitiveness by creating a more consistent identity that tests new markets or products; addressing declining profitability or consumer confidence. This signals a revolutionization of the need to be noticed or considered, requiring brand improvisation.

11 Conclusion

Brands from yesterday have a rich history and profound significance. Brand actors should avoid past brands when choosing whether to create, market, or launch an innovative brand. Selecting which specific brand signal to keep and which to discard could be a safer course of action. The business must be adamant about pushing the brand forward, but not everything will succeed, and new signals will undoubtedly need to be created. The preferred signal resolve, when spoken smoothly and continuously, communicates a powerful story that aims to pique listeners' interest. Customers tend to be people who experience a great deal of relief from consistency. Conversely, consistency does not mean using the same object all the time. Brands accomplish this by reiterating their position in the market. They go above and beyond to live up to their brand promise. One dependable improvement is the guarantee's longevity. In today's fast-paced world, the company would effectively regress and lose its innovative edge if it stood still brand or new product. No new product offering or brand can make an authentic claim to tradition. For heritage brands to be successful, they must look back, reclaim what made them great, rebuild the foundation and move forward.

References

1. Keller, K. L. (2007). *Strategic brand management.* Pearson Prentice Hall.
2. Kotler, P., & Keller, K. L. (2007). *Marketing management.* Pearson Prentice Hall.
3. Ramasamy, V. S., & Namakumari, S. (2007). *Marketing management.* Macmillan Business Books.
4. Zeithamal, V. A., & Bither, M. J. (2007). *Service marketing.* Tata McGrawhill.
5. Belch, G. E., & Belch, M. A. (2007). *Advertising and promotion.* Tata McGrawhill.
6. Schiffman, L. G., & Kanuk, L. L. (2007). *Consumer behaviour.* Pearson Prentice Hall.

Digital Marketing and Customer Satisfaction in the Hotel Industry

C. Kathiravan and P. Prabu

Abstract In the digital era, the hotel industry faces the same challenge as other industries: reaching and engaging with customers without the use of digital marketing is nearly impossible. To increase bookings, encourage repeat business, and enhance guest satisfaction, it has become crucial for hotels to market their amenities and services online. Therefore, the objective of this chapter is to explore the relationship between digital marketing and satisfaction of hotel guests. To achieve this goal, a combination of surveys conducted among hotel visitors and in-depth interviews with hotel managers will be employed as research methodologies. By analyzing the data collected, this chapter aims to provide insights as to how hotel management and marketers can effectively use digital marketing strategies to improve guest satisfaction in Chennai.

The study will primarily focus on investigating how digital marketing practices within the hotel industry in Chennai impact customer satisfaction. Digital marketing will serve as the independent variable and utilization, advantages and disadvantages, and perception as its sub-variables. On the other hand, customer satisfaction will serve as the dependent variable, encompassing sub-variables such as consumer hotel booking decisions and overall satisfaction. To gather data for this descriptive study, a questionnaire will be distributed to 404 hotel guests in the Chennai area.

In conducting the analysis, multiple regression analysis will be employed as the chosen method. By utilizing this statistical technique, the study aims to establish a comprehensive understanding of the relationship between digital marketing and customer satisfaction in the Chennai hotel industry. Based on the results obtained from the analysis, it can be concluded that digital marketing has a positive

C. Kathiravan (✉)
Department of Business Administration, Annamalai University, Deputation to Govt. Arts and Science College Sivakasi, Sivakasi, Tamil Nadu, India

P. Prabu
Department of Business Administration, Annamalai University, Deputation to Govt. Arts and Science College Sivakasi, Sivakasi, Tamil Nadu, India

Research Scholar, Department of Business Administration, Annamalai University, Working at Govt. Arts and Science College Sivakasi, Sivakasi, India

J. M. K. P. et al. (eds.), *Business Resilience and Digital Technology in the Post-Pandemic Era*, Contributions to Environmental Sciences & Innovative Business Technology, https://doi.org/10.1007/978-3-031-48075-1_9

correlation with customer satisfaction and holds significant influence within the Chennai hotel industry.

Keywords Digital marketing · Customer satisfaction · Hotel industry

1 Introduction

The hospitality industry has long been known for its fierce competition, as hotels constantly seek innovative ways to attract and retain guests. In today's digital age, hotels must embrace the power of digital marketing to effectively reach their target audience, considering the substantial amount of time people spend online. By leveraging various online platforms such as social media, email, search engines, and mobile apps, hotels can establish valuable connections with potential guests. This cost-effective approach not only enhances brand visibility but also drives an increase in bookings, making digital marketing an indispensable tool for the thriving hospitality sector.

To remain competitive in today's digital landscape, hotels must invest in digital marketing and make use of data analytics. Building a dedicated customer base, enhancing the hotel's online reputation, and increasing revenue are all possible, thanks to effective digital marketing.

Customer satisfaction, defined as the level to which consumers are satisfied with a company's products and services, has received increased focus in recent years. According to Hom [1], the service provider's ability to exert influence over the customer's experience is diminished or eliminated when the customer is actively engaged in the delivery process. Among the most distinguishing characteristics of a service are its intangibility, inseparability, heterogeneity, and perishability. Both Abdolali Khatibi Akbar and Pei Mey Lau contributed to its 2005 publication. Due to the intangible nature of services, it may be challenging for a business to grasp how customers evaluate and rank the quality of such services. Therefore, the service's quality is based on the accuracy of the user's input. The quality of service may differ greatly depending on the supplier, the customer, and the time of day.

2 Review of Literature

In a study conducted by J. K. Ayeh, N. Au, and R. Law in 2018, an extensive examination was carried out to analyze how digital marketing influences hotel performance. The researchers explored various digital marketing strategies, such as search engine optimization (SEO) and social media marketing (SMM), within the hospitality industry. Their findings revealed that incorporating these tactics can significantly affect the financial success of a hotel.

Kaur and Singh, in their literature study of the impact of digital marketing on retaining customers (2019), analyze the connection between digital marketing and repeat business in hotels. The authors discovered that digital marketing's ability to improve customer engagement, brand awareness, and loyalty was due in large part to its ability to facilitate more specific and individual forms of communication.

T. G. helps businesses get a better read on consumer happiness. In his work, Vavra [2] provides detailed recommendations. In this paper, the author discusses in depth the following aspects of quality function deployment: sample/customer participant selection, questionnaire design, interview/survey administration, data analysis, and action plan preparation. The author contends that if we had a model of happiness, it would shed light on the chain of events leading up to and including happiness. The suggested model consists of three interconnected phases: preconditions, satisfaction growth, and outcomes. Due to its memory-holding capacity, past experience is often cited as the best predictor of future happiness. Some people believe that the pleasure process includes evaluating the actual experience in relation to one's expectations. One's expectations are their complete belief in how strongly a product or service will display or deliver a given trait. In the provided model, wishes serve as an example of how expectations grow. Elangovan [3] examines environmentally conscious consumer behavior regarding eco-friendly packaging, contributing insights to environmental marketing in the hotel industry's context. Prendergast and Thompson [4] delve into the origins of cynical segmentation in environmental markets, probing whether it arises from marketers' attitudes or corporate directives. Kandampully and Suhartanto [5] investigate how customer satisfaction and image contribute to nurturing enduring customer loyalty within the dynamic hotel industry landscape.

According to Willard Hom, there are two main types of customer satisfaction models: macro-models, which are broad in scope, and micro-models, which drill down into more nuanced aspects of customer satisfaction. In addition, non-conventional methods of measuring customer happiness are compared with conventional ones in this study of marketing research. The marketing research literature review conducted for this study focuses on macro-models, such as "value," "quality," "complaining behavior," and "loyalty," and micro-models, which describe qualities like disappointment, equity, attribution, emotion, and regret.

3 Research Methodology

3.1 Research Design

The investigation's goals and emphasis guide the research design utilized in this study. The researcher employs a descriptive research approach to conduct the investigation. Descriptive research aims to depict the qualities and attributes of individuals or groups. The research design provides a portrayal of the current state of affairs. Focusing on Chennai, the study analyzes digital marketing and customer

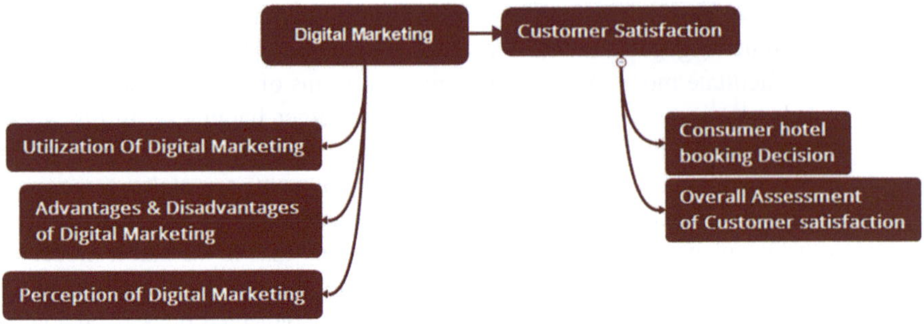

Fig. 1 Framework of the research

satisfaction within the hotel industry. The objective is to assess the existing levels of digital marketing and satisfaction among hotels in Chennai (Fig. 1).

3.2 Objectives of the Study

- Investigate the state of DM in the hotel business in Chennai
- Determine the level of contentment experienced by hotel patrons when staying in Chennai
- Analyze how different hotels in Chennai view the relationship between digital marketing success, digital marketing, and guest satisfaction
- Learn if and how DM influences their happiness

3.3 Hypotheses of the Study

- There is no clear demographic profile-based difference in DM among hotels in Chennai.
- Customer satisfaction in Chennai's hotel industry does not differ noticeably by guest's demographic makeup.
- Minimal discussion surrounds the efficacy of direct marketing, digital marketing, and customer satisfaction in Chennai's hotel industry.
- DM has minimal impact on customer happiness.

3.4 Data Collection

A well-structured questionnaire was used to collect primary data from hotel guests, with the first section covering the demographics of the staff, the second covering digital marketing, and the third covering guests' levels of satisfaction with their stay.

3.5 Reliability Analysis

To assess the reliability of this criterion, a study was carried out using a range of 0.80–0.85, where a satisfaction value of 0.70 was considered acceptable.

S. no.	Variable	Item	Cronbach's alpha
I	Digital marketing	39	0.85
II	Customer satisfaction	21	0.80

3.6 Sampling Technique

The research employed a sample process probability model, and the selection of questionnaire recipients was based on simple random sampling. This data collection method was both uncomplicated and cost-effective. By utilizing an equal chance approach, qualified customers in the hotel industry were identified. This particular survey methodology was designed specifically for this study to ensure the inclusion of hotel industry consumers.

3.7 Sample Size

Initially, a total of 438 questionnaires were distributed. However, 13 of them were never received, resulting in 425 returned questionnaires. Out of these, only 404 were considered usable due to serious issues with the remaining samples. Therefore, the final sample size consisted of 404 participants.

3.8 Data Analysis

For the purpose of this study, digital marketing was considered an independent variable, while customer satisfaction was treated as the dependent variable. The relationship between these two concepts was examined using path analysis.

4 Research Methodology

We conducted a survey in the construction industry to gather feedback from customers. Initially, we reached out to 438 customers and asked them to participate in the survey. However, only 404 individuals responded and provided their comments.

To assess the reliability of each survey section, we utilized Cronbach's alpha and employed SPSS (Statistical Package for the Social Sciences) for both factor analysis and Cronbach's alpha calculations. For evaluating the measurement and structural models, we employed the SmartPLS method. This method ensures rapid feedback and verifies the convergence and discriminatory validity of the measures. By employing SmartPLS, we were able to validate our metric and structural models, allowing us to explore the impact of digital marketing and changes in online consumer behavior on clients' satisfaction with built environments.

Digital marketing in this context encompasses three distinct facets: "Utilization," "Advantages and Disadvantages," and "Perception." These dimensions are further broken down into sub-variables, including consumer hotel booking decisions and overall customer satisfaction. To analyze the model's building blocks and their evolution, we used principal component analysis within the PLS framework, as suggested by Chin [6]. PLS has been advocated by Chin et al. [7] as a valuable analytical approach for enhancing accuracy.

The structural model was evaluated by estimating the path coefficients that connect the model's components and assessing the fit of the model to the data. First, we evaluated the measurement model, followed by the testing of the structural model. The fit of the model to the data was assessed by comparing the construct measures, such as validity and reliability (Fig. 2).

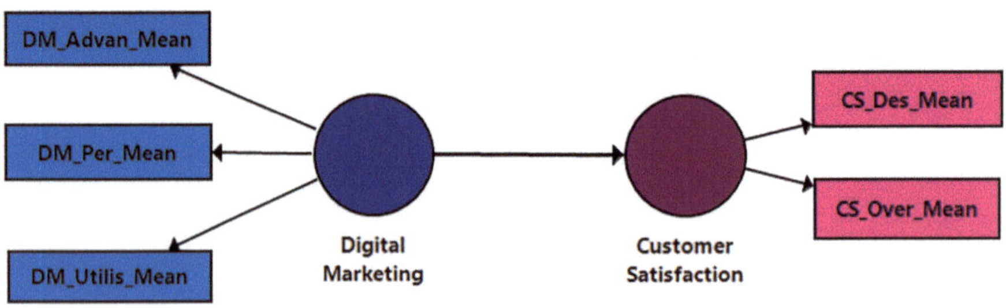

Fig. 2 Structural model

4.1 *Results and Discussions*

4.1.1 Measurement Model

We conducted tests using SmartPLS 2.0.M3 to assess the quality of our model, following the criteria proposed by Tenenhaus et al. [8]. These criteria involve analyzing the structural model as separate layers: the measurement model, the structural model, and the structural regression equation.

To evaluate the reliability of the sub-factors, we utilized SmartPLS software to assess the factor loadings. We adopted a minimum factor loading of 0.45, based on the recommendation of, to identify sub-factors. Additionally, we considered loadings above 0.50, as suggested by Hulland [9], to determine the reliability of the individual sub-factors.

4.1.2 Reliability

For assessing construct reliability and internal consistency, we employed composite reliability rather than Cronbach's alpha in PLS-SEM. Composite reliability provides a more suitable measure, as it avoids the assumption of equal consistency among all indicators made by Cronbach's alpha.

4.1.3 Convergence

Table 1 summarizes our evaluation of convergent validity for the credible constructs. We utilized SmartPLS software, along with the AVE test, composite reliability scores, and Cronbach's alpha [10]. All measurements of internal consistency reliability met or surpassed the threshold of 0.7, as recommended by Litwin [11]. This indicates that the dimension sub-factor appropriately represents the respective constructs.

Table 1 Factor loading for indicators of latent constructs

S. No	Factors and sub-factors	Factor loading	Cronbach's alpha	Composite reliability	AVE
1	Digital marketing		0.834	0.967	0.798
	Utilization of digital marketing	0.714			
	Advantages and disadvantages of digital marketing	0.876			
	Perception of digital marketing	0.834			
2	Customer satisfaction		0.854	0.967	0.687
	Consumer hotel booking decision	0.987			
	Overall assessment of customer satisfaction	0.808			

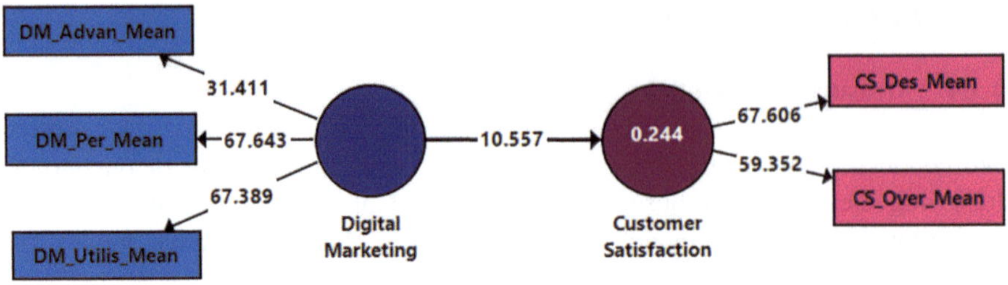

Fig. 3 Showing the bootstrapping diagram

Table 2 Path coefficients along with their bootstrap values and "T" values

Factors	Original sample (O)	Sample mean (M)	Standard deviation (STDEV)	T statistics	Supported	Significance values
Digital marketing –> customer satisfaction	0.738	0.660	0.042	13.027	Yes	$p < 0.01$ 0.000**

4.1.4 Structural Model Analysis

We employed a bootstrapping re-sampling approach [12] with 380 samples to assess the level of consequence of the pathways established within our model. Figure 3 of the structural model, which was validated through our research, illustrates this analysis. The SmartPLS tool was utilized for the examination (Table 2).

- Supporting and significant findings from the initial sample indicate that customer satisfaction has a direct and positive effect on digital marketing ($\beta = 0.738$, $t = 13.027$, $p = 0.01$).

The data presented here shows that there is a 0.738 significant reduction of one unit in OHS digital marketing and customer satisfaction.

5 Suggestions

- In the given context, "consumer satisfaction" pertains to the internal and external experiences and reactions of buyers throughout their search for and purchase of a product or service. It also encompasses the satisfaction that sellers can observe regarding these experiences and reactions.
- Customer satisfaction levels and their changes serve as valuable indicators for comprehending the reasons behind customers altering their opinions about a product. This is particularly beneficial for individuals seeking to predict profit, choice, repurchase behavior, or market share based on customer satisfaction.

6 Conclusion

This study delves into the previously unexplored realm of customer satisfaction and happiness regarding hotel stays in Chennai. Geissler [13] has established a strong correlation between digital marketing and customer satisfaction. The hotel industry has become inseparable from digital marketing, thanks to the remarkable influence of technology and the Internet on how people search for and reserve accommodations. In Chennai, this research marks the pioneering effort in investigating the connection between digital marketing and customer satisfaction within the hotel industry. Undoubtedly, digital marketing has brought about a revolution in this sector, presenting fresh prospects for growth and profitability. Hotels that proactively adopt digital marketing strategies and embrace technological advancements are poised to reap immense benefits from this paradigm shift.

References

1. Hom, W. (2000). *An overview of customer satisfaction models*. RP Group Proceedings, California Community Colleges.
2. Vavra, T. G. (1997). *Improving your measurement of customer satisfaction: A guide to creating, conducting, analysing, and reporting customer satisfaction measurement programs*. American Society for Quality.
3. Elangovan, A. (2006). Environmental marketing: Ecologically concerned consumers and the consumerist behaviour of environmentally friendly packaging. Doctoral Dissertation, Pondicherry University.
4. Prendergast, G. P., & Thompson, E. R. (1998). Cynical segmentation of environmental markets: The product of marketers' dispositions or corporate demands? *Journal of Euro Marketing, 6*(4), 17–34.
5. Kandampully, J., & Suhartanto, D. (2000). Customer loyalty in hotel industry: The role of customer satisfaction and image. *International Journal of Contemporary Hospitality Management. Bradford, 12*(6), 346.
6. Chin, W. W. (1998). The partial least squares approach to structural equation modeling. In G. A. Marcoulides (Ed.), *Modern methods for business research* (pp. 295–336). Lawrence Erlbaum Associates.
7. Chin, W. W., Marcolin, B. L., & Newsted, P. R. (2003). A partial least squares latent variable modeling approach for measuring interaction effects: Results from a Monte Carlo simulation study and an electronic-mail emotion/adoption study. *Information Systems Research, 14*(2), 189–217.
8. Tenenhaus, M., Vinzi, V. E., Chatelin, Y. M., & Lauro, C. (2005). PLS path modeling. *Computational Statistics & Data Analysis, 48*(1), 159–205.
9. Hulland, J. (1999). Use of partial least squares (PLS) in strategic management research: A review of four recent studies. *Strategic Management Journal, 20*(2), 195–204.
10. Fornell, C., & Larcker, D. F. (1981). Evaluating structural equation models with unobservable variables and measurement error. *Journal of Marketing Research, 18*(1), 39–50.

11. Litwin, M. S. (1995). *How to measure survey reliability and validity*. Sage Publications.
12. Efron, B., & Tibshirani, R. J. (1993). *An introduction to the bootstrap*. Chapman & Hall/CRC.
13. Geissler, G. L. (2002). Designing and maintaining an effective web site to facilitate online customer relationships. In S. M. Broniarczyk & K. Nakamoto (Eds.), *NA-advances in consumer research* (Vol. 29, pp. 496–497). Association for Consumer Research.

Paradigm Shift in Digital Transformation for Sustainable Growth in Indian Healthcare: A Proposed Theme-Based Exploratory Study

R. K. Prema and B. Arul Senthil

Abstract Today's business environment heavily relies on data, which are helpful for the growth of a business. Manufactures are launching products that are based on customers' needs and wants. Digital platforms are serving this purpose for both customers and producers, including those in the healthcare industry. Lately, consumers have been more willing to purchase medical products online. Although they encounter challenges in the supply chain, the industry is moving toward more online delivery. To accomplish this, the healthcare industry is increasingly working with new technology.

This chapter contains reviews from journal articles, websites, and news articles on topics related to healthcare advancement, and its findings are also based on data obtained from various secondary sources. The study is exploratory and descriptive in nature. An extensive literature review was conducted to support the proposed research objectives, and the results are discussed. The sustainable growth of a country not only lies in providing healthcare facilities but also in reducing the death rate, in this case by increasing the use of digital delivery.

Keywords Healthcare · Technology · Digital transformation · Sustainability

1 Background

Health is a required service for every human being, irrespective of income. It is one of the unavoidable services that we all need. Around the globe, the COVID-19 pandemic threated human health and well-being. The world has become virtual, and its citizens are giving less preference to physical interactions and consultations with

R. K. Prema
Acharya Bangalore B-School, Bangalore, Karnataka, India

B. A. Senthil (✉)
Jansons School of Business (Autonomous), Coimbatore, Tamil Nadu, India
e-mail: arulsenthil@jsb.ac.in

J. M. K. P. et al. (eds.), *Business Resilience and Digital Technology in the Post-Pandemic Era*, Contributions to Environmental Sciences & Innovative Business Technology, https://doi.org/10.1007/978-3-031-48075-1_10

healthcare providers. When the nature of the patient's journey changes, the service providers must follow suit. India's digital connectivity is expected to grow at a rate of 80% by 2034, which is six times greater than that of 2014. Rural Internet users are expected to increase by 58% annually [1]. This projection depicts an expansion in digital access to healthcare services and may transform the entire business in the healthcare sector. As more and more patients use Android smartphones, the healthcare sector's digital opportunities likewise increase. In modern society, people do not have time to wait, and they require services to be available at their fingertips and at any time. *Digital health* is not a new term. Before digitization, patients used to consult their physicians over the phone or in person to seek treatment. Nowadays, because the integration of technology has been increasing, there is greater scope for digitally delivering health services. Income patterns, population, customer-centric business practices, and COVID-19 impacts were the prime reasons for transforming healthcare into a digital space. This business has to deal with uncertainty, whereas data play vital roles in decision-making and information delivery. The key research aim in this chapter is to focus on understanding the drivers of digital adoption, digital intervention, and investments in healthcare and their impacts on quality care.

1.1 Introduction

Digital transformation in healthcare sounds overwhelming, but it has a positive impact on healthcare. Organizations are facing hectic pressures to automate and digitalize [2]. Businesses have adapted to recent technologies to serve their customers' needs and to remain competitive.

Researchers have identified several theories to understand adopting new technologies for different motives, from 1964 to 2014 [3]. The theory of reasoned action (TRA) [4] and the theory of planned behavior [5, 6] are the important theories in understanding customer and patient behavior. The digital transformation in healthcare has focused on connecting healthcare professionals with patients by considering time, the quality of services, consulting, awareness, resource optimization, and knowledge transmission. One of the drivers of the adoption of technology is self-service technology in healthcare. The service delivery in healthcare depends on the value of services available in the channel flow between people, which is mediated by the availability of various technologies [7, 8]. Customers' being more active and aware benefits service providers [9]. Information and communication technology is used in every sector, and the healthcare sector is no exception. Digitalization's pathway multiplies the number of access points that serve the public [10]. Though technology improves patient access to healthcare, a comprehensive patient-centric approach is required to make it successful. Internet service providers and users of healthcare must work together in digital spaces [11].

Digital transformation in healthcare aims to solve problems by prompting noteworthy changes to information gathering, computer programs, communication channels, and technologies [12]. Because new technologies are constantly being introduced, the changes have to be frequently updated [13, 14]. Digital transformation requires technical support to facilitate the core organizational process and to change business models. The ever-increasing population means that healthcare demands tend to increase along with the emergence and spread of novel diseases. Public and private healthcare often cannot meet the in-person needs of patients, because of the surplus of patients accessing healthcare. However, guidance can be given online to reduce the number of inpatient visits at hospitals. The workload of healthcare staff is increasing because of the increasing number of patients and the development of telemedicine. The integration of technology with healthcare marks shifts in how healthcare professionals work, the use of IT, and training competent healthcare staff [15]. Companies must also adapt to by introducing and adopting new business models. E-health platforms can connect most distant patients with physicians. The increasing administrative burden in healthcare over the past few decades has reduced physicians' quality time spent with patients [16, 17].

2 Research Methodology and Framework

The research methodology adopted for the study was qualitative. The researcher followed three phases to substantiate the method of this research.

1. Phase I involved identifying relevant studies and conducting a literature review. The literature search was carried out on electronic databases such as CINAHL, PubMed, Scopus, Elsevier, Sage, WoS, PsycINFO, ScienceDirect, ABI/INFORM, SpringerLink, and NCBI; also, Google scholar was used as a supplement. The researcher included studies from 2010 to 2020 and used the following keywords: "digital transformation," "digitalization," "digital health," "digital healthcare," "e-health," "healthcare industry," "digital adoption in healthcare," "digital components in healthcare," and "digital challenges." The researcher reviewed the papers on the basis of the objectives proposed in each study. In total, more than 90 reviews were collected from electronic databases—out of which, that that closely fit within the context of this study were streamlined.
2. Phase 2 involved categorizing the literature review according to this study's objectives. Tables 1, 2, and 3 display the collected literature reviews and their links to themes.
3. Phase 3 involved an analysis of the literature review. In the final stage, the literature review was interpreted, analyzed, and discussed.

Table 1 Drivers of Digital Adoption

Authors	Focus	Subtheme	Findings
Atluri et al. [18]	The extensive research in this study measured the impacts of digital technologies on health providers' insurers and consumers.	Tech-enabled consumers and healthcare landscape.	There is an increased use of digital technologies in healthcare among these players.
Keesara et al. [19]	The article focused on justifying the response to the COVID-19 outbreak and its relationship with the digital revolution in healthcare.	The surge of COVID-19 and technological outbreak.	The article brings insight to the necessity for quick action as a responsive measure to the pandemic and the need for digital tools in healthcare.
Ford et al. [20]	This chapter focused on identifying the digital disruption role and innovation in healthcare.	Business model services and innovation in healthcare.	The findings provide insights on how digitization in healthcare impacts the value chain across the sector.
Currie and Finnegan [21]	This study attempted to identify the association between the policy and the practice of introducing an ICT program in the UK NHS.	Adoption of EHR in the UK NHS.	The study inferred that electronic health records inform healthcare clinicians, managers, and patients and transform the implementation process.
Westbrook and Braithwaite [22]	This study focused on measuring automation in clinical activities and its impact on efficiency and safety.	ICT and its disruption on the healthcare system.	Embracing ICT-disruptive technologies leads to an innovative model of healthcare delivery.

3 Results and Discussion

Because the healthcare sector has less and less time for in-person care, digitally delivering healthcare services has become increasingly important. Digital adoption helps healthcare professionals to identify diseases early and treat them well in advance. The increased population in India and the availability of devices on the healthcare market provide opportunities for the sector to be even more digitalized. Indian healthcare evolves as an excellent platform for investments. As the demand for healthcare increases, there will be more investment in this field, which may accelerate the industry's performance. Digitalization in healthcare has a noticeable impact on unemployment in the country because of its high demand and adoption of technology.

3.1 Theme 1: Drivers of Digital Adoption

Digitalization brings in technology-enabled solutions that could reduce burdens on healthcare infrastructure and inpatient stays. Around the globe, digitalization transformation has been profound and extensively distributed through the business world.

Table 2 Digital Intervention and Investments in Healthcare

Authors	Focus	Subtheme	Findings
Laurenza et al. [23]	This research reviewed the literature on how Information Technology (IT) influences business growth in healthcare sector.	Analysis on the digital adoption in a healthcare case study.	Digital technology adoption in healthcare minimizes workloads and simplifies the complex process. It also increases efficiency and quality care.
Moro Visconti and Quirici [24]	The aim of this study was to assess the economic benefits of digitalization.	Digital technology and healthcare investment.	This study concluded that digital initiatives add value to the healthcare setting and hence promote investment opportunities.
Sehgal and Hooda [25]	The aim of this study was to assess the position and predictions of the made in India initiative.	Made in India for healthcare.	It concludes that the initiative was useful, and its use depends on its formulation, development, and implementation by the government.
PwC [1]	The report aims to present and discuss the Indian healthcare sector in terms of population, funding, and associated factors.	Report on funding Indian healthcare.	It addressed the need for increased spending on healthcare in India and the country's deficiencies and population growth.

Communication in healthcare was once limited, before the wide adoption of technology in medicine. Thanks to the availability of online information, e-health applications, and wearable devices, the number of people adopting digital health services has increased [18]. The nature of work and business strategies are changing in healthcare, reflecting provider-centric models. This has forced patients and physicians to also shift toward digitalization [20–22]. The introduction of e-health technologies in this field offers excellent e-health applications, thereby connecting the provider with the customer/patient [37]. The advantage of e-health is that real-time communication is more effective because of the increase in the number of people who are carrying a mobile phone wherever they go. Gig-economy workers also help to meet the demands of the healthcare sector. The pandemic drove the digitalization of products and the technological revolution in healthcare, which significantly improved the digital delivery of health services [19]. The public rely on technological progress not only to access healthcare services but also to make their decisions. Extensive research has shed light on doctors' services, the availability of medical facilities, treatment options, and costs (Chart 1).

Table 3 Digital Intervention and Quality Care

Authors	Focus	Subtheme	Findings
Alami et al. [10]	The study investigated the ability of the health system to compete with the business trends concerning Information and Communication Technology (ICT).	Digital health and challenges.	The new technological shifts inevitably give birth to a parallel health system for improve effectiveness and efficiency.
Herrmann et al. [26]	The research motive of this study was to understand the technological enablers around digital and disruptive innovation in healthcare.	Digital transformation in the healthcare sector.	Digitalization brings effectiveness and optimizes cost. The organization has to look into the investment opportunities in this sector.
Mohanan et al. [27]	This paper highlighted the quality of care on public and private players and the challenges required to improve the quality of care.	Healthcare quality's challenges and priorities.	Participation from state and central government is required to achieve quality care. Policymaking and capacity building need to be prioritized.
Kim et al. [28]	This study aimed to understand how the institutional structure facilitates information sharing and impacts outcomes.	The healthcare industry's process of sharing knowledge.	Knowledge sharing was strongly influenced by the organizational infrastructure and also enhanced patient safety.
Lenz et al. [29]	This research looked into IT-integrated services, focusing on quality improvement and cost minimization.	Healthcare process's achievements and challenges.	The results recorded dimensions such as flexibility, awareness, process modeling, validation, verification, and evaluations of patient services.
Buntin et al. [30]	This study focused on determining health IT and its effects on outcomes.	Review on benefits of health IT.	The pros of information technology on health are evolving, and IT moving healthcare toward a paradigm shift in the implementation process.
Herrmann et al. [26]	This study's aimed to identify what types of digital innovations are emerging with digital transformation.	Digital transformation and disruption on healthcare.	Identified that business models can be strengthened by the innovative transformation that brings flexibility in healthcare.
Desmedt et al. [31]	This study focused on assessing the role of IT in delivering care.	Whether health IT improves care.	The findings showed that information technology in healthcare helps integrate care but that this can happen only with well-equipped state-of-the-art technologies.
Cavallone and Palumbo [32]	This study focused on the applications and drawbacks of industry 4.0 in healthcare.	Myths of industry 4.0.	Industry 4.0 provides a platform for the effectiveness of healthcare through artificial intelligence. Issues also exist.

(continued)

Table 3 (continued)

Authors	Focus	Subtheme	Findings
Herrmann et al. [26]	This research focused on measuring the e-health intervention in treating health problems.	Health professional acceptance and adoption of e-health.	E-health literacy was assessed. The acceptance of e-health was low, but still, acceptance of online aftercare was moderate.
Helleso et al. [33]	This research explored the shift of patient information from hospital to home care and its effectiveness.	Digitalization of patient information.	There was a difference between the patients transferred from hospitals to those transferred home care, concerning the characteristics of ICT.
Alami et al. [10]	The research aimed to understand the ability of the healthcare system to cope with the new, digital transformation.	Digital health and challenges in transformation.	It found that India must be resourceful in supporting the new initiative, to reap the benefits.
Labovitz et al. [34]	This study focused on the use of artificial intelligence (AI) in minimizing risk on the mobile platform with stroke patients.	Using AI to reduce patient risk.	Patients who used a smartphone enjoy the benefits of the potential increase in real-time monitoring and service quality.
McGrail et al. [35]	This empirical study aimed to identify the impacts of virtual visits on the overall quality of care.	Virtual visits and patient-centered care.	Online visits benefit patients and benefited them more when they used a known service provider.
Mehra and Sharma [36]	This study aimed to measure policies, practices, and research in the healthcare industry.	Sustainability measures.	To achieve sustainable healthcare, industry needs to promote green healthcare initiatives and reduce waste.

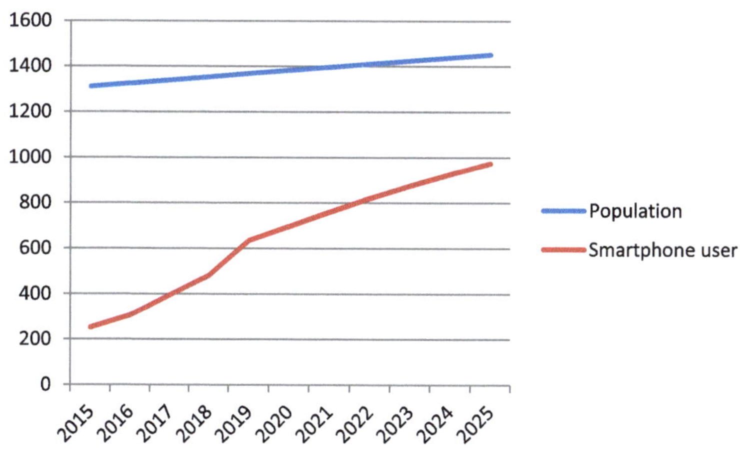

Chart 1 Comparison of Population and Number of Smartphone user globally. Source: Data were taken from Statista X-axis: year; Y-axis: values in millions

3.2 Theme 2: Digital Intervention and Investments in Healthcare

Health is one of any country's social and economic investments because it accelerates growth [23]. The outbreak of the COVID-19 pandemic and the upcoming year's funding for Indian healthcare requires a USD 300 billion investment to prepare for unprecedented events. The recent innovation in technology in general and the COVID pandemic has brought numerous changes to the healthcare sector in delivering services. For example, a few digital components, such as telemedicine, wearable devices, artificial intelligence, robotics, and blockchain technology have reshaped the entire industry. Spending on health is an investment, not just an expense. Increasing healthy life years has positive effects on a country's economic growth. The healthcare sector in India is widening and still in high demand. Health impacts a country's economic growth directly through labor productivity, and therefore, health is considered an investment good. The COVID-19 pandemic has alerted the public to the need for investment in their health.

Healthcare sector demands have triggered private players' participation; they deliver around 60% of inpatient and 80% of outpatient care to the country [25], encouraging Foreign Direct Investment (FDI) in this sector. The growth of venture capital and private equity investments and the success of Initial Public Offering (IPO) and National Health Protection Scheme (NHPS) Public-Private Partnership (PPP) have increased investment in healthcare from such global healthcare players [24]. Digital platforms add value to the health ecosystem, promoting investment and pay-for-performance for the sake of cost optimization. The public is more concerned about their care, thereby appreciating insurance claims as sources of investment into their health. As technology drives growth, there has been an increase in involvement and active participation from private equity players and an increase in FDI [1].

3.3 Theme 3: Digital Intervention and Quality Care

"India is home to global leaders in innovation in and quality of health care" [27]. Technology has advanced, and the average life span is increasing compared with earlier years [23]. Recent technological innovations in medicine, clinical technology, and information technology have balanced hospitals' costs and quality performances [38]. Digital technologies accurately capture vital statistics, which helps bring about greater compliance and more-effective coordination. It also pipelines the process flow in hospitals through predictive analytics. In line with the findings of Li and Collier [38], CT directly impacts clinical quality, whereas Information Technology (IT) affects the entire process's quality. The process and organizational performances in healthcare depend on information and knowledge sharing [28, 29]. Therefore, for knowledge sharing, this technological breakthrough acts as a platform and improves the quality of care. Any flaws in the medical process would affect outcomes. More precisely, medical processes must be well planned, organized, documented, communicated, evaluated, and available for future reference [29]. Coordination from allied industries is required to deliver quality care, and essential solutions lie in digital technology and its comprehensive utilization [30]. The decision support system in healthcare enhances patient safety and impacts costs.

Digitalization helps in the collection and analysis of health data. Digital transformation improves healthcare performance and quality of care by reducing costs [26]. Desmedt et al. [31] promote integrated care, which needs support from information technology. Patients will be more autonomous and be able to practice self-management as society is experiencing a shift from an anonymous society to a post-COVID-19 digital world, which puts organizations under stress.

Industry 4.0 allows physical and virtual resources to combine and deliver reliable outputs. With the evolution of digital platforms and the increased use of web-based platforms, patients' needs are being better met. We also found a slight increase in the outcome of service delivery [39].

Cavallone and Palumbo [32] investigated the application of Industry 4.0 in healthcare in terms of tools and devices and found that it enhances the administration and management of resource availability and focuses on strengthening patient–provider relationships. The aim of the study was to determine whether such technology reduces medical errors, saves time, and increases service quality. Digitalization helps healthcare organizations to promptly deal with complexity and unpredictability. One of the facilities in Electronic Medical Record (EMR) is to record the medical history of patients which can be safely shared among hospitals. Hence, patients are placed at the central of health services and undergoing continuous care [33]. Digital tools improve patient health. Digitalization maintains stronger connections among stakeholders, clients, clinicians, patients, and physicians in highly challenging times. Digital health is a keystone revolution in healthcare systems, increasing efficiency and public health effectiveness [10]. In line with the findings of Porter and Teisberg [40] and Christensen et al. [41], because of the increasing demand of an aging population and rising costs, healthcare systems face significant challenges. Despite the availability of high-tech devices in the United States, the average cost is higher than the quality care delivered, and this disparity is appearing on a global scale [42]. Digital technology interventions can minimize these challenges, thus enhancing healthcare delivery [34, 35]. Moreover, India's population impacts the quality of healthcare delivered because it has too few resources for its population size. Digitalization in healthcare helps to balance the doctor–patient ratio in the delivery of quality care. This study's literature review attempts to understand the revolution in healthcare resulting from digitalization. The results are not conclusive, though, because the researcher is in the process of analyzing the secondary data.

4 Limitations and Implications for Future Research

The study suffers from certain limitations. The research is qualitative, and it is a review paper, and therefore, the researcher cannot substantiate new insights. Though the study follows a well-established objective, the researcher could not fetch reviews on the themes of digitalization and investment. Recent reviews over the past 10 years were considered for the research themes because the transformation happened within this gap. Digitalization impacts different clinical fields differently. Those areas were not discussed in-depth; only the surface area was covered. There are directions for future research which can help future researchers. First, future research can focus on assessing the digitalization in healthcare by administering a questionnaire to providers and patients to better understand how it integrates into business models. Second, digitalization brings business transformations. To reap the benefits, researchers can try to determine which supports or government initiatives can be focused on. Third, future research can focus on changes in government policies and the reachability of services for the public. Fourth, researchers can focus on population and service delivery with the help of IT services.

5 Conclusion

In recent times, business organizations have been looking for strategies that focus on service innovations [43]. If we need to upscale digital health, interventions must be made from all the stakeholders, especially the government. Digitalization in healthcare enhances the overall consumer experience and delivers the needs of all healthcare consumers. It empowers healthcare workers to perform their best, improving the quality of care. Patients, the end users in this sector, are empowered by digital transformations and are more involved in decision-making. Technology is bringing every one of us much closer, but effectively utilizing it matters. Medical care focuses not only on treating diseased but also on prevention and personalized care. Hence, integrating the services through an efficient platform with technological integration better serves society.

The healthcare industry is going through a digital revolution. The need to provide patients with high-quality care and top-notch experiences is paramount. The new norm is to provide personalized treatment in this strong, competitive landscape and to make the industry operationally efficient. Transparency and patient privacy have become the two most critical constituents. Cloud infrastructure, big data, artificial intelligence, and blockchain are the key technologies that have brought a wave of change in the healthcare industry. Health for all through digitalization is the new agenda in India. In addition to this, the pandemic opens up opportunities for digital health and innovation diagnostics. The prevailing challenge in healthcare is that access, affordability, and healthcare quality impact life expectancy. Digitalization in this sector accelerates the improvements in healthcare outcomes, enabling easier access to quality healthcare infrastructure. There has been a shift from inpatient care to outpatient services through technology-enabled services, thereby balancing the population and staffing ratio. Disruptive trends are revolutionizing the health industry today. A few of those trends include a tremendous shift in demographics and social change, economic power, globalization, and technology-driven and -empowered customers. More customers are empowered now, and hospitals are approaching accreditation bodies to enhance their quality standards and benchmarks. To increase economic output, citizens must be healthy. The reachability of the health service depends on the extent to which it effectively uses technology. Hence, this research concludes that a well-established digital health service not only provides quality care but also enhances the health status of the country, thereby contributing to an increase in economic wealth. A digital initiative in healthcare has significant impacts on the quality of care and leads to sustainable growth.

References

1. PwC. (2016). Indian healthcare on the cusp of a digital transformation. https://www.pwc.in/assets/pdfs/publications/2016/indian-healthcare-on-the-cusp-of-a-digital-transformation.pdf
2. Prema, R. K., & Kathiravan, C. (2020). An investigative study on role of HR analytics for predictive business decision in 4.0. *International Journal of Scientific Research in Science, Engineering and Technology (IJSRSET), 5*(11).
3. Almeida, J. P. L. d., Farias, J. S., & Carvalho, H. S. (2017). Drivers of the technology adoption in healthcare. *BBR. Brazilian Business Review, 14*(3), 336–351.
4. Fishbein, M., & Ajzen, I. (1975). *Belief, attitude, intention and behavior: An introduction to theory and research*. Addison-Wesley.
5. Ajzen, I. (1985). From intentions to actions: A theory of planned behavior. In J. Kuhl & J. Beckmann (Eds.), *Action control: From cognition to behavior* (pp. 11–39). Springer Verlag.
6. Ajzen, I. (1991). The theory of planned behavior. *Organizational Behavior and Human Decision Processes, 50*, 179–211.
7. Akaka, M. A., Vargo, S. L., & Schau, H. J. (2015). The context of experience. *Journal of Service Management, 26*(2), 206–223.
8. Fliess, S., Dyck, S., & Schmelter, M. (2014). Mirror, mirror on the wall – How customers perceive their contribution to service provision. *Journal of Service Management, 25*(4), 433.
9. Immonen, M., & Koivuniemi, J. (2018). Self-service technologies in health-care: Exploring drivers for adoption. *Computers in Human Behavior, 88*, 18–27.
10. Alami, H., Gagnon, M. P., & Fortin, J. P. (2017). Digital health and the challenge of health systems transformation. *Mhealth, 3*, 31.
11. Zylberberg, T., & Mamou-Mani, G. (2017). *Les industriels du numérique en santé prêts à cosigner un pacte de responsabilité avec les acteurs publics*. TIC-Santé.
12. Vial, G. (2019). Understanding digital transformation: A review and a research agenda. *The Journal of Strategic Information Systems: Incorporating International Information Systems, 28*, 118–144.
13. Hwang, J., & Christensen, C. M. (2008). Disruptive innovation in health care delivery: A framework for business-model innovation. *Health Affairs (Millwood), 27*(5), 1329–1335.
14. Perakslis, E. D. (2017). Strategies for delivering value from digital technology transformation. *Nature Reviews. Drug Discovery, 16*(2), 71–72.

15. Sturesson, L., & Groth, K. (2018). Effects of the digital transformation: Qualitative study on the disturbances and limitations of using video visits in outpatient care. *Journal of Medical Internet Research, 20*(6), e221.
16. Seto, R., Inoue, T., & Tsumura, H. (2014). Clinical documentation improvement for outpatients by implementing electronic medical records. *Studies in Health Technology and Informatics, 201*, 102–107.
17. Vollmer, A.-M., Prokosch, H.-U., & Bürkle, T. (2014). Identifying barriers for implementation of computer based nursing documentation. *Studies in Health Technology and Informatics, 201*, 94–101.
18. Atluri, V., Cordina, J., Mango, P., Rao, S., & Velamoor, S. (2016, August). *How tech-enabled consumers are reordering the healthcare landscape*. McKinsey & Company.
19. Keesara, S., Jonas, A., & Schulman, K. (2020). Covid-19 and health care's digital revolution. *New England Journal of Medicine, 382*(23), e82.
20. Ford, G., Compton, M., Millett, G., & Tzortzis, A. (2017). The role of digital disruption in healthcare service innovation. In M. A. Pfannstiel & C. Rasche (Eds.), *Service business model innovation in healthcare and hospital management* (pp. 57–70). Springer International Publishing.
21. Currie, W. L., & Finnegan, D. J. (2011). The policy-practice nexus of electronic health records adoption in the UK NHS: An institutional analysis. *Journal of Enterprise Information Management, 24*(2), 146–170.
22. Westbrook, J. I., & Braithwaite, J. (2010). Will information and communication technology disrupt the health system and deliver on its promise? *Medical Journal of Australia, 193*(7), 399–400.
23. Laurenza, E., Quintano, M., Schiavone, F., & Vrontis, D. (2018). The effect of digital technologies adoption in healthcare industry: A case based analysis. *Business Process Management Journal, 24*, 1124.
24. Moro Visconti, R., & Quirici, M. C. (2020). Digital technology and efficiency gains in healthcare infrastructural investments. *Healthcare Research and Public Safety Journal, 2*(1), 1–8.
25. Sehgal, S., & Hooda, S. (2015). *Emerging role of private sector in indian health care delivery market: Trends, pattern and implications*. Intern report, Institute for Studies in Industrial Development (ISID).
26. Herrmann, M., Boehme, P., Mondritzki, T., Ehlers, J. P., Kavadias, S., & Truebel, H. (2018). Digital transformation and disruption of the health care sector: Internet-based observational study. *Journal of Medical Internet Research, 20*(3), e104.
27. Mohanan, M., Hay, K., & Mor, N. (2016). Quality of health care in India: Challenges, priorities, and the road ahead. *Health Affairs, 35*(10), 1753–1758.
28. Kim, Y. M., Newby-Bennett, D., & Song, H. J. (2012). Knowledge sharing and institutionalism in the healthcare industry. *Journal of Knowledge Management, 16*(3), 480–494.
29. Lenz, R., Peleg, M., & Reichert, M. (2012). Healthcare process support: Achievements, challenges, current research. *International Journal of Knowledge-Based Organizations, 2*(4), 52–68.
30. Buntin, M. B., Burke, M. F., Hoaglin, M. C., & Blumenthal, D. (2011). The benefits of health information technology: A review of the recent literature shows predominantly positive results. *Health Affairs, 30*, 464–471.
31. Desmedt, M., Pless, S., Dessers, E., & Vandijck, D. (2017). Integrating and safeguarding care: The potential role of health information technologies. *International Journal of Care Coordination, 20*(3), 112–118.
32. Cavallone, M., & Palumbo, R. (2020). Debunking the myth of industry 4.0 in healthcare: Insights from a systematic literature review. *The TQM Journal, 32*(4), 849–868.
33. Helleso, R., Ekstedt, M., & Sockolow, P. (2018). Digitalization of patient information process from hospital to community (home) care nurses: International perspectives. *Studies in Health Technology and Informatics, 250*, 227.

34. Labovitz, D. L., Shafner, L., Reyes, G. M., Virmani, D., & Hanina, A. (2017). Using artificial intelligence to reduce the risk of nonadherence in patients on anticoagulation therapy. *Stroke, 48*(5), 1416–1419.
35. McGrail, K. M., Ahuja, M. A., & Leaver, C. A. (2017). Virtual visits and patient-centered care: Results of a patient survey and observational study. *Journal of Medical Internet Research, 19*(5), e177.
36. Mehra, R., & Sharma, M. K. (2021). Measures of sustainability in healthcare. *Sustainability Analytics and Modeling, 1*, 100001.
37. Meneghetti, A. (2013). Challenges and benefits in a mobile medical world: Institutions should create a set of BYOD guidelines that foster mobile device usage. *Health Management Technology, 34*(2), 6–7.
38. Li, L. X., & Collier, D. A. (2000). The role of technology and quality on hospital financial performance. *International Journal of Service Industry Management, 11*(3), 202–224.
39. Hennemann, S., Beutel, M. E., & Zwerenz, R. (2017). Ready for eHealth? Health professionals' acceptance and adoption of eHealth interventions in inpatient routine care. *Journal of Health Communication, 22*(3), 274–284.
40. Porter, M., & Teisberg, E. (2006). *Redefining health care: Creating value-based competition on results*. Harvard Business, School Press.
41. Christensen, C. M., Bohmer, R., & Kenagy, J. (2000). Will disruptive innovations cure health care? *Harvard Business Review, 78*(5), 102–112.
42. Porter, M. E., & Teisberg, E. O. (2004). Redefining competition in health care. *Harvard Business Review, 82*(6), 64–76.
43. Kim, M., Song, J., & Triche, J. (2015). Toward an integrated framework for innovation in service: A resource-based view and dynamic capabilities approach. *Information Systems Frontiers, 17*(3), 533–546.

A Study on the Financial Performance of Arbion Infra Services Limited

A. A. Bhavya and S. Jayashree

Abstract Financial statements are important in determining the company's financial position. They help compare the company's performance with different periods and companies in the same industry. With the help of these comparisons, the management of the company is able to take various decisions on behalf of the company. This research study focuses on the comparison of financial statements of Arbion Infra Services Limited, Kerala, in the last 5 years, 2018–2022. The comparison is made using the current, efficiency, profitability ratios, and trend analysis. With the help of the findings, various conclusions and suggestions were drawn. The study has concluded stating that the company must focus on reducing its expenses and effectively using its resources.

Keywords Financial statements · Profitability ratios · Trend analysis · Financial position

1 Introduction

The business world is dynamic in nature, and companies need to keep up with this to survive in the competitive market. The main factor for a business to continue its operation is finance and the efficient and effective management of the finance.

The word "performance" means "to do." It means to execute an action or to fulfill an activity. So financial performance means the extent to which the financial objectives of an organization have been achieved by the company. It is a tool used to measure an organization's overall development and growth.

A. A. Bhavya
Kristu Jayanti College (Autonomous), Bengaluru, India

S. Jayashree (✉)
Department of Professional Accounting and Finance, Kristu Jayanti College (Autonomous), Bengaluru, India
e-mail: jayashree.s@kristujayanti.com

J. M. K. P. et al. (eds.), *Business Resilience and Digital Technology in the Post-Pandemic Era*, Contributions to Environmental Sciences & Innovative Business Technology, https://doi.org/10.1007/978-3-031-48075-1_11

A detailed way assessing a company's financial strength can happen through the financial performance analysis. Investigating a company's financial records, reports, assets, and processes is a part of it. Financial analysis is conducted to ascertain whether a company is steady, solvent, liquid, or profitable enough to justify an investment in the economy. This is carried out by examining the financial statements of the company, like balance sheet, profit and loss account, cash flow statements, and statement of changes in equity.

Financial performance analysis includes observing the policies and business operations in addition to looking at the financial statements and measures. As a result, tracking financial performance is now more important than ever in the business world. The growing importance of financial performance analysis is outlined below.

1.1 Trend Analysis

Utilizing trend data that has recently been identified, the technical analysis process is known as trend analysis attempts to anticipate future variations in stock prices. Trend analysis, which also uses data such as price movements and transaction volume, can be used to anticipate the protracted course of market volatility.

2 Review of Literature

- Eljelly [1] in the research on "Liquidity-Profitability trade-off: An empirical investigation in an emerging market" investigated the relationship between profitability and liquidity, as calculated by the current ratio and cash gap on a sample of public companies in Saudi Arabia. The research found a notable inverse relation between the company's profitability and liquidity level, as measured by the current ratio.
- Singh [2] and Pandey [3] proposed that for the efficient functioning of any business firm, fixed and current assets play an important role and that the use of working capital is crucial as it has a straight influence on profitability and liquidity. They examined the working capital components and identified a notable effect of working capital management on profitability for Hindalco Industries Limited.
- Saluja and Kaur [4] conducted research on the "profitability performance of public sector banks in India" on nationalized banks for an interval of 10 years from 1997–1998 to 2006–2007. They found out the component accountable for the financial performance of the nationalized banks.
- Tiwari and Parray [5] elucidated the financial statement analysis of Ranbaxy ltd. They provided a detailed understanding of two highly used financial techniques, ratio analysis and usual size statement analysis. The purpose of the research was

to help the users comprehend how these techniques should be put to use to understand the financial position of a company.

3 Statement of the Problem

The profitability position varies on the liquidity and leverage of the companies. In this regard, this study makes an attempt in analyzing the financial data to find the financial performance of Arbion Infra Services Ltd.; therefore, this study has been undertaken.

4 Objectives of the Study

(a) To analyze the financial statement and to find out Arbion Infra Services Ltd.'s financial position.
(b) To find out the liquidity and operating efficiency of Arbion Infra Services Ltd.
(c) To analyze the solvency ratios and examine the ability of Arbion Infra Services Ltd. to pay off its debts.

5 Research Design and Methodology

The research design used in the study is the descriptive type, which enables the researchers to describe the phenomena as reported by the researchers' information. Only quantitative data has been used for the performance of the study.

5.1 Type of Data Collected

The data used for the study is secondary in nature. The research uses the secondary data provided by Arbion Infra Services Ltd. The information was collected from the office and their annual reports.

5.2 Trend Analysis

Trend analysis is a way of looking at data from long-term studies over time, comparing results from quizzes and tests or comparing datasets from regularly distributed satisfaction surveys. It helps identify trends.

5.3 *Trend Analysis of Sales*

Analyzing sales trends involves looking at historical data to find patterns. This is a helpful technique for budgeting and financial analysis, which can indicate when a company's short-term revenue growth rate is starting to shift.

6 Limitations of the Study

1. The data used for the study is purely quantitative; therefore, the qualitative aspect of the company is ignored.
2. As the study is entirely dependent on secondary data, the results and findings are purely proportional to the accuracy of such data.
3. The study is conducted based on the information provided by the company.

7 Analysis

7.1 *Trend Analysis of Net Profit*

The net profit trend analysis's overall direction of change for the period under consideration shows an increase in profitability. The consistent pattern demonstrates that while profitability has not decreased, there has also been little advancement in the revenue over the course of time (Tables 1 and 2).

Table 1 Depicting trend analysis of sales

Year	Sales (₹)	Trend %	Difference
2017–2018	17,61,82,687	100	0
2018–2019	27,33,44,674	155.15	55.15
2019–2020	30,23,16,344	110.6	10.6
2020–2021	29,56,23,726	97.79	−2.21
2021–2022	36,35,62,532	122.98	22.98

Table 2 Depicting trend analysis of net profit

Year	Net profit (₹)	Trend %	Difference
2017–2018	73,86,924	100	0
2018–2019	1,02,10,167	138.22	38.22
2019–2020	1,16,27,070	113.88	13.88
2020–2021	84,74,854	72.89	−27.11
2021–2022	1,03,39,321	122	22

7.2 *Trend Analysis of Total Assets*

Trend analysis of total assets is done by taking the total assets from the previous year's financial statement to analyze and compare it to the other years to understand the increase or decrease in the trend percentage (Table 3).

8 Findings

1. The sales trend had a drastic increase of 55.15% in 2018–2019, and it reduced for the next two periods to 97.79% in 2020–2021 and went on to increase by 22.98% in the year 2021–2022. This gives a conclusion on the revenue earned and the overall profitability of the firm.
2. The net profit trend increased by 38.22% in the year 2018–2019 and decreased in the period of 2020–2021 to 72.89%. A low net profit ratio reveals that the expenses of the firm have been more than that of its revenue.
3. The total assets trend had a drastic increase of 193.64% from the base of 100, after which there was a huge downfall in the subsequent years, which has continued to go on to decrease to 106.86 in 2021–2022. This shows that the company had high growth during the initial period of study, which gradually reduced, and as the assets increase, the liability and equity will also increase.

9 Summary of Findings

After conducting the financial statement analysis, it is clear that the company is in an acceptable position because:

- It has maintained a satisfactory level of current assets against its current liabilities.
- The company's performance in the year 2020–2021 has not been great as it is the year in which the company earned the least amount of profit.
- The company has been paying off its debts on time, without exceeding the credit period allowed.
- They have sufficiently good control in place in relation to collection of the receivable amount, and amount is collected back even though the collection period has exceeded the limit.

Table 3 Depicting trend analysis of total assets

Year	Total assests (₹)	Trend %	Difference
2017–2018	2,66,18,327	100	0
2018–2019	7,81,61,520	293.64	193.64
2019–2020	11,78,03,494	150.72	50.72
2020–2021	11,58,37,033	98.33	−1.67
2021–2022	12,37,83,975	106.86	6.86

- The gross profit has been high in most of the years, but net profits have reduced, showing that the company has incurred more of expenses resulting in the low operating profit.

10 Conclusion

The company's financial position is an important factor for sustaining in the market in the long run as well as for the overall development of the company to move forward and expand in the future. The study conducted on the financial position gives a conclusion that the company has a satisfactory level of liquidity and solvency position. The profitability position of the company has above the ideal standard in the case of ROCE and gross profit, while operating and net profit has been reduced after the period of 2019–2020. It is found that the company has been up to date in maintaining its receivables and in paying off its liabilities on time.

Due to the COVID-19 pandemic, the performance of the company in terms of liquidity, profitably, and solvency has been affected, leading to a low performance in the recent years. Policies and procedures must be undertaken by the company to overcome this situation and to ensure that it continues to make profit in the future, to sustain in the competitive market for a foreseeable period of time.

References

1. Eljelly, A. M. (2004). Liquidity-profitability trade-off: An empirical investigation in an emerging market. *International Journal of Commerce and Management, 14*, 48–61.
2. Singh, P. (2008). Inventory and working capital management: An empirical analysis. *The Icfai University Journal of Accounting Research, 35*, 40–45.
3. Pandey, S. (2008). Impact of working capital management in the profitability of Hindalco Industries Limited. *The ICFAI University Journal of Financial Economics, 36*, 10–20.
4. Saluja, J., & Kaur, R. (2010). Profitability performance of public sector banks in India. *Indian Journal of Finance, 4*(4), 17–25.
5. Tiwari, A., & Parray, F. S. (2012). Analysis of short-term financial position – A case study of Ranbaxy ltd. *ArthPrabhand: A Journal of Economics and Management, 1*(6), 36–50.

Websites

6. https://arbioninfra.com/.
7. https://www.wallstreetmojo.com/.
8. www.investopedia.com.
9. www.wikipedia.com.

Driving India's Sustainable Future: The Rise of Green Economy and Green Jobs

Grace Blessy

Abstract The green economy is a new economic paradigm that aims to promote economic growth while safeguarding the environment and promoting sustainable social and economic development. This necessitates a shift to green careers and skills and the creation of new occupations associated with workforce greening. All major industries, including renewable energy, construction, manufacturing, construction, transportation, and agriculture, can benefit from green jobs. For all facets of education, training, and enterprises, skill acquisition and improvement have significant positive effects. A low-carbon, resource-efficient, and socially inclusive economy is what is meant by the term. In a green economy, infrastructure and assets that enable decreased carbon emissions and pollution, improved energy and resource efficiency, and the prevention of the loss of biodiversity and ecosystem services generate growth in employment and income. To achieve sustainable development, social inclusion, and a shift to a green economy, according to the ILO/UNEP report "Working towards Sustainable Development—Opportunities for Decent Employment and Social Inclusion in a Green Economy", is essential. The report, which was released nearly 4 years after the first study by the Green Jobs Initiative, lays out a wealth of policy lessons, best practices, and successful programmes and shows that a green economy with more and better jobs, poverty reduction, and social inclusion is both necessary and achievable.

Keywords Green jobs · Green economy · Sustainability · Technology · Industry

1 Introduction

Sustainable development faces a problem, an opportunity, and a requirement in the shift to a green economy, a necessity in the sense that current resource use and emissions from production and consumption go beyond the carrying capacity of

G. Blessy (✉)
Kristu Jayanti College (Autonomous), Bangalore, India

J. M. K. P. et al. (eds.), *Business Resilience and Digital Technology in the Post-Pandemic Era*, Contributions to Environmental Sciences & Innovative Business Technology, https://doi.org/10.1007/978-3-031-48075-1_12

natural systems. Although the consequences and scarcities are unevenly distributed from place to region, it is impossible to ignore the misuse of land, soil, water, and other natural resources and the loss of biodiversity. Similarly, emissions of greenhouse gases, pollutants, and garbage exceed the capacity of the ecosystem to regenerate. The situation worsens as populations and incomes rise in many emerging nations. Productivity and consumption do not align with a sustainable economy or the planet's limits.

2 Objectives

- To analyse the ideas and implementation strategies for the employment effects of green economy policy
- To develop green economy strategies and employment promotions

3 Methodology

The study aims to identify the green economy that encourages people and nature to coexist peacefully and tries to meet all of their needs concurrently. Environmental policies, sustainable agriculture, alternative energy sources, and wildlife conservation are some of the topics that green economics may research.

4 Green Economy Concept

In recent years, various stakeholders have created various ideas of what makes up a green economy. There must be harmony between environmental sustainability and economic development, notwithstanding their divergent priorities. There are macroeconomic ideas that aim to measure and enhance the environmental performance of the entire economy, as well as ideas that concentrate on creating a market for green products or eco-friendly technologies, instruments of economic and environmental policy for organizing and directing the shift to a green economy.

Several policy tools have been considered to encourage the transition to a green economy. While markets, or supply, demand, and the essential infrastructures for environmentally friendly items, are being built, a green economy should emerge. For this objective, a variety of market-based tools are suggested. Market-based instruments are designed to place a price on emissions and the use of natural resources or to raise this price and provide incentives for environmental protection. A second category of market-creating policy instruments is also considered in this analysis. These include laws and other initiatives that encourage the purchase of eco-friendly goods and services. Supply-side instruments, or actions to promote technology and investment as part of an all-encompassing industrial and innovation policy, are typically used to supplement these.

A sustainable infrastructure policy is another component for advancing a green economy. Long-term choices are made regarding the development of public infrastructure, and these choices significantly impact how economic structures and the technology they support develop [1].

Last, a successful green economy policy needs a constantly shifting instrument mix to adjust to shifting market and technological advancements. To do this, monitoring and evaluation systems are crucial components of green economy initiatives.

Ideas and Implementation Strategies for the Employment Effects of a Green Economy Policy.

The literature on the effects of environmental policy on innovation, markets, and employment has produced the following significant findings:

- While evaluating investment initiatives, the impacts of short-term replacement are frequently examined.
- Long-term innovation impacts are increasingly being considered in studies on the consequences of environmental policy regulation, and in this setting, progressively favourable interrelated effects are being found.
- The possibility of a "double dividend" with concurrently lower social security contributions is covered in the research on the economic impacts of environmental tax reform. It is abundantly evident from the various methodologies and findings that evaluating the employment consequences of specific policy instruments is impossible without considering other relevant influences, most notably a country's capacities. As a result, it is impossible to quantify the effect of any one instrument on employment alone; rather, it depends on factors such as market conditions, the availability of trained labour, the work requirements of technology alternatives, etc. The analysis of the expected action mechanisms in the specific context, along with the identification of the pertinent influencing elements, is a prerequisite for analysing the impact on employment.
- Compared to traditional technologies, there are signs that many environmentally friendly technologies are linked to higher labour intensity and, as a result, boost employment. Some technologies are predicted to have higher capital intensity, which will have a detrimental impact on employment.
- Energy price hikes have minimal effects on employment unless labour expenses are reduced with money from the energy tax. Positive results are anticipated in this situation.
- Rigidities in the labour market can postpone the benefits of employment.
- Environmental policy's impact on innovation is a contentious issue. Opportunity costs limit the innovation that can be done, although environmental advances are also encouraged.
- Environmentally friendly technologies can help nations set the bar for environmental policies to prosper economically in international trade.

- Nonetheless, it is currently thought that there is little chance that polluting industries will move to nations with laxer environmental standards.
- The management of natural resources and the valuing of ecosystem services both have employment possibilities.

5 Green Economy Strategies and Employment Promotion

The transition to a green economy is associated with economic structural change, affecting labour markets. To facilitate this fundamental shift, labour markets must be created to be dynamic and effective. Additionally, several unique difficulties for employment policy in the context of a change to a green economy have been recognized [2].

Creating and adjusting education and training programmes to meet the skill requirements of the green industries is a significant problem. Dealing with contracting industries and the resulting job losses presents another difficulty in implementing green economy plans. They can involve, for instance, providing mediation services, further training for the impacted employees, or developing public employment services.

6 Developing Green Economy Strategies

Policy tools for the transition to a green economy cannot simply be applied from one nation to another; they must be customized to the local economic, institutional, and social circumstances. So examining the relevant value chains and the potential for establishing a local supply chain is a crucial first step in assessing various alternatives for raising labour demand in the framework of a green strategy. When enacting environmental taxes or other types of regulation that affect the cost of resources or energy, consideration should also be given to the current economic structures.

Moreover, opportunities for innovation that support employment and economic growth can be established in critical environmental policy action areas. Environmental problems in other nations may have economically viable answers that lead to export success. This, however, can only be anticipated by concentrating certain resources on areas of activity that are particularly relevant in the home country. Sacrifice and restrictions are not always part of an economy focused on sustainably protecting and utilizing the environment.

On the contrary, several effective examples demonstrate how an environmental focus and prospects for economic development can coexist.

These are the causes:

1. The costs that can be avoided due to environmental degradation

   Direct expenses increase as the environment deteriorates in many nations. Climate change and water scarcity already impact developing countries with sizeable agricultural sectors. Yet the costs associated with soil, water, and air pollution are also economically significant and can be at least partially avoided by focusing on the environment.
2. Environmental technologies and eco-friendly business strategies presenting economic opportunities

   The demand for environmental technologies is increasing, according to numerous research. In these markets, emerging nations now account for a more significant portion. New services that use fewer resources and the environment have been added. New business models for selling environmental technologies and offering services tailored to local contexts exhibit significant growth potential, particularly in developing nations.
3. Potential of natural resource management for economic development

   Several studies demonstrate that valuing natural resources in the framework of sustainable management techniques can spark a dynamic that can balance economic development with the preservation of nature and the environment.

The expectation of increased employment in fields based on resource conservation and sustainable resource management is linked to a green economy.

This hope is fuelled by the new employment generated in environmental technologies and renewable energy sources in Germany and other Western industrial nations.

From the Expansion of Environmental Industries to the Effects of Green Growth on Employment.

The influence of a green economy on employment can be theoretically defined and restricted in various ways, given the ongoing international debate on green growth and the economy. The phrase "green jobs" refers, in other words, to the sectoral understanding of a green economy, to jobs that are created in the context of the growth of a green economy in developing environmental industries, to improve environmental impact, or to ensure compliance with environmental standards in established sectors (such as environmental management). This economic sector's expansion is typically solely considered for its implications on gross employment.

6.1 Enhancing Employment Policy Skills for a Green Economy

Governments must decide which industries should be subject to new environmental restrictions or which sectors should be prioritized when creating nation-specific green economy initiatives. This entails, among other things, evaluating how well each technological discipline can advance the national economy and its impact on employment [3].

The accountable authorities, however, frequently lack the information required to be able to make such choices. Governments should refrain from intervening specifically in economic processes; the conclusion was frequently reached in the past. In the framework of a green economy policy, this potential is not a choice, though. A crucial tool for creating a green economy is developing new ecologically friendly technology and the targeted burdening of already-existing filthy businesses, both of which call for commensurate State requirements. When deciding not to implement any environmental policy interventions in specific sectors, drawing up a definition with an influence on employment and the sectoral distribution of economic activity in the nation is required.

7 Green Job Concept

An output and a process perspective can be distinguished to differentiate between green jobs and regular employment. From a production standpoint, "green jobs" refer to employment opportunities in businesses and industries that make either wholly or largely environmentally friendly products. From a process standpoint, the term extends beyond this and includes work that aims to lessen the environmental effect of businesses that do not, in some way, create environmental commodities.

The following are some guidelines for creating a quantifiable definition of green jobs:

- Concepts and terminology must be closely related to eco-friendly business practices. Starting with economic activity, the definition of a green job identifies employment associated with that activity.
- For the planned data collection, definitions relating to the environment must be applicable.
- When measuring occupational employment, the observational unit is the job.
- Using recognized statistical/classification techniques, data should be organized by industry (kind of good or service produced) and occupation (type of job performed).
- The concept of "green jobs" should theoretically recognize three elements: jobs involved in manufacturing green products, greening production processes, and being part of the supply chain for producing green goods and services. Techniques and the extent to which the factors above can be measured are still up for debate and require further research.
- Inclusion of elements like pay or benefits would be inappropriate; the definition should only be based on economic activity.

Green jobs are often positions where the work is directly tied to producing or delivering green goods or services associated with green economic activity categories. The most frequently used words refer to environmental and energy-related challenges, such as energy efficiency, alternative energy use, conservation, and cleanup. While defining the green economy, these shared components are also considered.

8 Green Jobs as Decent Jobs

Many international organizations are debating the effects of a green economy and green jobs on the employment standard. So, it is necessary for green occupations also to be respectable jobs. Since the late 1990s, the ILO has promoted the idea of decent work, which is supported by four pillars:

1. Employment advancement and its placement at the centre of economic and social policy
2. The expansion of social security to cover all employees and their families
3. The implementation of fundamental labour standards.
4. Social discourse

The mutual strengthening of both concerns and an improved ability to reach an agreement should result from proving the fundamental compatibility of an environmental outlook and upholding social values. Yet the simultaneous tightening of the definition of "green jobs" means that the employment share in green industries is no longer accurately reflected.

9 Green Jobs and Their Impact on the Economy

Green jobs are described by the United Nations Environment Programme (UNEP) as "positions in agricultural, manufacturing, R&D, administrative, and service industries targeted at substantially preserving or restoring environmental quality." In other words, environmental jobs focus on preserving and enhancing the environment or constantly consider and work to minimize their impact on the ecosystem's health. The International Labour Organization (ILO) classifies this kind of work as follows:

- Increases effective raw material and energy use
- Cuts back on carbon gas emissions
- Cuts down on waste and contamination
- Conserves and rebuilds ecosystems
- Helps with climate change adaptability

These green occupations also have a positive impact on the world economy. The ILO has issued a warning that, in the absence of any improvements, future job growth will fall short of meeting the demands of the expanding workforce in rising and developing nations. The Global Employment and Social Outlook 2018 report from this organization states that "changes in production and use of energy to attain the two °C target may contribute to the creation of about 18 million jobs in the world economy." More extensive energy use from renewable sources, the rise of electric vehicles, and construction work to achieve energy efficiency in buildings are just a few of the changes that will be made to comply with the Paris Agreement and create green jobs [4].

10 India Has a Green Economy

The global agenda for sustainable development now includes the "green economy" as a core theme. The fast growth of India over the past 10 years has increased employment opportunities and raised living standards. However, due to the environment's degradation and the depletion of natural resources, its remarkable growth record has been constrained. As a result, significant steps have been taken to achieve a green and decarbonized economy. With customers now focused on a greener economy, thanks to COVID-19, brands embrace sustainability by default. India must therefore shift to a circular economy with the help of the authorities and businesses.

Urbanization is a worldwide phenomenon but accelerates in developing nations like India. According to a United Nations projection, 60% of the world's population will reside in cities by 2030. Ninety percent of the world's rural population is currently found in Asia. Nonetheless, the area is experiencing a rapid urbanization rate, predicted to reach 56% by 2050.

Developing nations like India can completely change their economies by taking advantage of the opportunities presented by urbanization, primarily driven by population growth and faster industrialization. The climate is changing significantly due to this increase in urbanization though.

Pollution levels in the air, water, and soil are rising, and this is due to urban areas. The detrimental repercussions of over-urbanization include excessive groundwater depletion brought on by overdevelopment and poor management, excessive carbon emissions from city-dwelling automobiles, spatial congestion brought on by urban sprawl, and so forth. Large Indian cities are experiencing a population boom that is dangerous for the environment and the climate, as well as placing a significant strain on the nation's infrastructure for managing electricity, water, and transportation [5].

11 India's Status as a "Green Economy"

The 2020 Environmental Performance Index ranks nations based on waste management, air quality, biodiversity and habitat, fisheries, ecosystem services, and climate change.

India placed 169th out of 180 nations among the top six major economies, showing that its green growth is behind. One of the economies in the world with the quickest growth rates is India. It is Asia's third-largest economy and the sixth-largest economy in the world by GDP. IMF estimates that COVID-19 caused a significant global economic contraction in 2020 but that macroeconomic recovery will cause growth of 6.0% in 2021 and 4.9% in 2022. In comparison to the same period in the previous year, India's GDP rose at a record rate of 20.1%, or 32.38 lakh crore, between April and June of 2021. The Indian economy is expected to grow by 8.3% and 7.5%, respectively, in 2021 and 2022, according to the World Bank [6].

As a result of the severe depletion of natural resources, including minerals, water, and fossil fuels, which would raise the cost of fuel, energy, and raw materials, expansion could have extremely negative environmental effects. The scope of India's green growth will depend on its capacity to gradually decrease reliance on the resources required to support economic growth while enhancing social fairness and generating jobs. In balancing these demands, green growth can be pretty important. The management of public debt and fiscal deficits, the two fundamental obstacles to national policymaking, could impede the technological advancements necessary for green growth. Moreover, macroeconomic policy would heavily depend on the trade balance. To optimize the development advantages of green growth initiatives across essential sectors like energy, trade, and income, it is necessary to understand how they affect each area.

12 Governmental Green Energy Initiatives

12.1 Hydrogen Energy Mission

The Hydrogen Energy Mission is a project that aims to produce hydrogen using environmentally friendly energy sources. This project has the potential to change the transportation industry completely. The usage of clean fuels in India would also be encouraged. The focus on green hydrogen in the budget aligns with the long-term objective of reducing reliance on batteries made of minerals and rare earth elements for energy storage and with the evolution of technology [7].

12.2 Public Transportation

For the first time, the cabinet has allocated private financing totalling INR 18,000 crores (USD 2.43 billion) for 20,000 buses, along with creative financing through public-private partnerships, which would fundamentally alter how India's public transportation system operates. The project seeks to lessen reliance on personal vehicles and, as a result, cut carbon emissions.

12.3 Deep Ocean Mission

The mission would carry out programmes to preserve deep-sea biodiversity in addition to deep ocean surveys and exploration. Within 5 years, this programme will receive a budget of over INR 4000 crore.

12.4 Urban Swachh Bharat Mission 2.0

With a focus on integrated management of manure, sludge, and sewage treatment, the classification of waste sources, the reduction of disposable plastics, and the reduction of air pollution, the government intends to manage waste from construction and demolition activities efficiently and bioremediate all inherited landfills.

12.5 Preference for Greener Goods Among Consumers

A recent survey reveals that the younger generation is aware of sustainable products. Customers favour purchasing goods from businesses that strongly emphasize waste reduction, carbon footprint reduction, sustainable packaging, dedication to ethical labour practices, and respect for human rights. Environmental awareness among people has grown as a result of the pandemic.

Consumers are now choosing recyclable plastic and fibre packaging to reduce environmental waste. When a company performs poorly on sustainability metrics, customers switch products or services, which creates market opportunities for competitors to innovate in favour of eco-friendly products.

Several FMCG companies have committed to sustainable development and chosen eco-friendly packaging options. The world's top 10 consumer goods manufacturers (Danone, Coca-Cola, Pepsi, Unilever, L'Oréal, etc.) set the audacious goal of attaining 100% sustainable packaging by 2025 in 2020 [8].

13 Booming Sectors with Green Job

Future professions will be developed due to the economy's decarbonization and the growth of the circular economy, and existing employees will be modified to fit the new, environmentally friendly reality. According to experts, the following industries have the most potential to generate green jobs.

13.1 Energy

The COVID-19 pandemic significantly impacted the global economy for the majority of 2020 and 2021, impacting both the volume and composition of energy consumption. The many forms of limitations placed in place, which strained supply chains and severely curtailed economic activity, have significantly impacted employment, notably in this area. The ninth edition of the "Renewable Energy and Employment Report: Annual Review 2021" predicts that despite this, the field of renewable energy will employ 12 million people in 2021, up from 11.5 million in 2019 [9].

13.2 Agriculture

Global organic food and drink sales climbed by 15%, according to research released in 2022 by the Organic Farming Research Institute (FiBL). The market increased to $129 billion in just 2 years, with revenues surpassing the $100 billion milestone in 2018. The pandemic's effects and a rise in consumer interest in organic food were both factors in the historic boom observed in 2020.

With a combined share of 90%, North America and Europe make up the majority of sales. Most of the expansion comes from other geographical areas, particularly Asia, as organic food markets have become prominent in nations like China, India, and South Korea [10].

13.3 Tourism

Ecotourism, also known as ecological travel, is a growing trend in a culture that is becoming more concerned with the state of the environment. This industry generates jobs in fields including adventure experience design, high mountain trail construction, protected area discoveries, and revitalization of rural communities at risk of extinction, such as eco-villages.

13.4 Transport

Almost 30% of the CO2 emissions in the European Union (EU) are attributable to this sector, with 72% of those emissions coming from the road transportation industry. Many nations have already taken action to lower transportation emissions significantly. By 2050, the EU will lower these by 60% compared to 1990 levels, generating job possibilities in electric automobiles, public transportation, and electrified railroad freight haulage.

Extensive research and recommendations on the economic benefits of investment in 11 green sectors:

- Renewables
- Garbage
- Businesses
- Forests
- Transportation
- Fishing
- Structures
- Agriculture
- Cities
- Tourism
- Water

What Type of Training Is Required?

There is no standardized training programme for what is known as "green collar workers" because so many green jobs are accessible today. Besides degrees, courses, and postgraduate courses with an ecological focus, environmental sector specialization constitutes green training for a particular position.

For instance, those who design packaging need to be knowledgeable about sustainable materials. An engineer who wants to work in the energy sector must specialize in everything related to renewable energy, energy efficiency, and the decarbonization of the Economy. A lawyer who cares about protecting the environment must specialize in environmental law.

14 Conclusion

In the context of the global discussion, the idea of a green economy has both a normative and an empirical component. Without considering the underlying compatibility of economic development and environmental sustainability, the macroeconomic conceptions of international organizations present several interpretations of a new concept of growth and the economy. In support of this, the sectoral notion of a green economy highlights prior achievements and the financial potential of the associated green economy strategy. The development and evolution of the environmental industry, environmental technologies, and environmental and economic activities are experimentally demonstrated by periodical investigations. They act as a foundation for justifying the development of certain industries and as a resource for further in-depth, sector-specific assessments. Making a meaningful line between the environmental sector and other sectors is challenging, which makes it tough to define and understand green jobs. In addition to employment in environmental sectors, there are workers in other industries whose goal is to improve environmental performance. The connection with ideas to guarantee or enhance the quality of employment leads to further difficulties. There is still no agreement on this area's operational requirements and accompanying metrics. National economic techniques aim to analyse the net employment effects of environmental policy changes and a structural shift to a green economy, in contrast to the sector-based ideas of environmental industries.

References

1. International Labour Organization. (2018). India: Transition to a greener economy: Country brief. Retrieved from https://www.ilo.org/wcmsp5/groups/public/%2D%2D-ed_emp/%2D%2D-emp_ent/documents/publication/wcms_615595.pdf
2. Government of India. (2022). Economic Survey 2022–23. Retrieved from https://economicoutlook2022.nic.in/uploads/ECONOMIC_SURVEY_2022-23.pdf
3. Ministry of Environment, Forest and Climate Change. (2020). India's Green Economy Report 2020. Retrieved from http://moef.gov.in/wp-content/uploads/2020/03/Green-Economy-Report.pdf
4. Niti Aayog. (2021). Green jobs in India: An overview. Retrieved from https://niti.gov.in/sites/default/files/2021-06/GreenJobsInIndia-Overview.pdf
5. The Energy and Resources Institute. (2022). Green jobs and sustainable development in India. Retrieved from https://www.teriin.org/sites/default/files/2022-05/GreenJobsSustainableDevelopmentIndia.pdf
6. Confederation of Indian Industry. (2019). Green jobs: Towards decent work in a sustainable, low-carbon world. Retrieved from https://www.greenjobs.net.in/pdf/Green%20Jobs%20Report.pdf
7. Indian Ministry of New and Renewable Energy. (2022). Annual report 2021–22. Retrieved from https://mnre.gov.in/sites/default/files/webform/notices/AnnualReport2021-22.pdf
8. Environmental Defense Fund. (2020). Green stimulus and recovery measures for India. Retrieved from https://www.edf.org/sites/default/files/content/green-stimulus-and-recovery-measures-for-india.pdf.
9. International Renewable Energy Agency. (2021). Renewable energy and jobs – Annual review 2021. Retrieved from https://www.irena.org/publications/2021/Apr/Renewable-Energy-and-Jobs-Annual-Review-2021
10. World Bank. (2021). India: Building back greener: Employment, skills, and economic recovery after COVID-19. Retrieved from https://openknowledge.worldbank.org/handle/10986/35408

An Influence of Neuromarketing on the Consumers of Small-Scale Industries in Varanasi District: An Explorative Study

Arti Singh, Ibha Rani, and Sanjeev Chauhan

Abstract Neuromarketing is the study of how the brain processes influence to consumer behaviors. It looks at what stimuli people react to and how their reactions inform marketing strategies. This includes understanding how people perceive products or ads, as well as how they respond to pricing, packaging, and promotional materials. As for its use in the real world, neuromarketing sounds like an interesting and worthwhile topic to explore consumer behavior. Hence, this chapter investigates the impact of neuromarketing on consumer behavior of small-scale businesses in Varanasi. Specifically, the chapter looked into how the use of neuromarketing strategies has impacted sales, customer satisfaction, and long-term customer engagement. For this purpose, the researcher has collected the primary data from 140 respondents and analyzed it statistically. Additionally, the challenges faced by these businesses in terms of implementing neuromarketing, as well as potential solutions or ways that neuromarketing, can be used more effectively by small businesses in Varanasi.

Keywords Neuromarketing · Consumer behavior · Customer satisfaction

1 Introduction

The area of discussion in this chapter is how neuromarketing has impacted the consumer behavior of small-scale industries in Varanasi. Neuromarketing is a multidisciplinary field that combines psychonomic, statistics, social and cognitive, and computer science to study individuals' brain waves and physiological responses to various stimuli. Neuromarketers have a significant impact on the world economy

A. Singh (✉) · I. Rani
Department of Commerce, Kristu Jayanti College (Autonomous), Bangalore, India

S. Chauhan
Institute of Business Management, GLA University, Mathura, India
e-mail: sanjeev.chauhan@gla.ac.in

J. M. K. P. et al. (eds.), *Business Resilience and Digital Technology in the Post-Pandemic Era*, Contributions to Environmental Sciences & Innovative Business Technology, https://doi.org/10.1007/978-3-031-48075-1_13

by affecting the price points of goods and services at lower levels, thereby reducing inflationary pressure. Neuromarketing makes use of post-scanning neuroimaging techniques to study the brain activation and physiological responses of individuals to various stimuli. This can be used in video games, advertisements, food packaging, website designs, and cinema tickets. It has been found that neuroimaging techniques could also be used to determine the need for certain goods and services.

This research is among a growing number of studies that are looking deeper into how neuromarketing strategies and technologies are affecting people's behavior. Neuromarketers in small-scale industries, such as hotels and restaurants, use these tools to attract more business. The study compared the impact of neuromarketing tools on the traditional marketing practices for international student travelers who visit Varanasi, which can be seen by reading the full article that talks about neuromarketing for Indian temples in particular.

1.1 Consumer Behavior in Small-Scale Industries in Varanasi

This chapter discusses the impact of neuromarketing on consumer behavior of small-scale industries in Varanasi. The author interviewed a marketing expert who mentioned that neuromarketing is the practice of adjusting marketing for a specific demographic, product, or market based on neurological studies about human behavior, online research, and computational modeling.

2 Literature Survey

The researcher has come across the following literature.

Singh and Singh's [1] article provides a comprehensive overview of consumer neuroscience and its implications for marketing research. The authors explain the concepts and techniques of neuromarketing, including the use of electroencephalography (EEG), functional magnetic resonance imaging (fMRI), and eye-tracking to understand consumers' subconscious responses to marketing stimuli. The article also discusses the advantages and limitations of neuromarketing, such as the need for specialized equipment, the difficulty in interpreting results, and ethical concerns.

Yadav and Arora [2] aimed to investigate the impact of neuromarketing on consumer behavior. The researchers conducted a survey using a structured questionnaire and collected data from 250 respondents. The results indicated that neuromarketing techniques significantly influence consumers' buying behavior, particularly in terms of their willingness to pay, preference for products, and brand loyalty. The study also revealed that gender and age differences play a significant role in the effectiveness of neuromarketing techniques. Overall, this study suggests that neuromarketing can be an effective tool for marketers to influence consumer behavior.

Srivastava and Singh [3] aimed to explore the emerging concept of neuromarketing and its application in marketing research. The authors concluded that neuromarketing has the potential to provide a deeper understanding of consumer behavior and preferences through the analysis of neural activity. It can also help in developing more effective marketing strategies and communication messages that resonate with consumers. However, the authors also highlighted the ethical concerns surrounding the use of neuromarketing and emphasized the need for proper guidelines and regulations to ensure consumer privacy and data protection. Overall, the study highlights the significance of neuromarketing as a tool for gaining insights into consumer behavior and improving marketing outcomes.

Chakraborty and Das [4] aimed to explore the emerging trend of neuromarketing in the field of marketing research. The researchers found that neuromarketing has gained significant attention from both academia and industry due to its potential to provide insights into consumer behavior that traditional research methods cannot. They also identified various techniques used in neuromarketing, such as EEG, fMRI, eye-tracking, and facial recognition, and their applications in different stages of the marketing process, including product development, branding, and advertising. Overall, the study concluded that neuromarketing has a promising future in the field of marketing research, but it requires further exploration and refinement.

Garg and Gupta [5] in their study on the role of neuromarketing in decision-making discuss the potential role of neuromarketing in decision-making. They explore how insights from neuroscience can be used to improve marketing strategies and decision-making. The authors argue that neuromarketing has the potential to provide more accurate and reliable data on consumer behavior, which can help marketers make better decisions. However, they also caution that ethical considerations need to be taken into account when using neuromarketing techniques. Overall, the study provides useful insights into the potential of neuromarketing to inform decision-making in marketing, but more empirical research is needed to fully evaluate its effectiveness.

Kumar [6] aimed to review the influence of neuromarketing on consumer buying behavior. The author found that neuromarketing techniques have a significant impact on consumer decision-making by tapping into their subconscious mind. The study further suggests that neuromarketing can be used to improve advertising and promotional strategies, product development, and customer experience. However, the author also highlights the ethical concerns associated with the use of neuromarketing techniques and the need for appropriate regulations in this field. Overall, the study suggests that neuromarketing has great potential in influencing consumer behavior and can provide valuable insights for marketers to improve their strategies.

Singh and Kharub [7] provide an overview of the concept of neuromarketing, its techniques, and its applications. The authors highlight the potential benefits of using neuromarketing in marketing research, such as understanding consumer behavior, improving product design, and increasing sales. They also discuss the ethical implications of using neuromarketing, particularly with regard to consumer privacy and autonomy.

Tiwari and Khanna [8] explored the impact of neuromarketing on consumer behavior through a conceptual approach. The authors reviewed the literature on neuromarketing and its various techniques, such as fMRI, EEG, and eye-tracking, and discussed their potential implications on consumer behavior. They concluded that neuromarketing can provide deeper insights into consumer behavior by identifying the unconscious responses of consumers to marketing stimuli. The study emphasized the importance of combining traditional research methods with neuromarketing techniques to develop effective marketing strategies. Overall, the authors suggested that neuromarketing has the potential to revolutionize the field of marketing research and provide more accurate insights into consumer behavior.

Fugate set out to cover the history of the field, explain the methodology in simple terms, and offer potential future directions for consumer behavior research. In a study titled "Neuromarketing: A Layman's Look at Neuroscience and Its Potential Application to Marketing Practice," he also hypothesized that combining neural activity with traditional tools could result in more effective marketing strategy.

da Rocha et al. validated EEG technology as a suitable neuromarketing tool and demonstrated that consumers' assessments of their satisfaction with aesthetic dermatological treatments involved the activation of neural circuits related to facial beauty assessment. In a study titled "A Neuromarketing Study of Consumer Satisfaction," EEG technology serves as a suitable neuromarketing tool and demonstrates how consumers' assessments of their happiness with cosmetic dermatological treatments entailed the activation of neural circuits related to assessments of facial appearance. This result is consistent with the left hemisphere's greater contribution to the volunteer's improved self-perception of appearance following treatment compared to before, as well as the high level of satisfaction with the volunteer's new appearance among family and friends.

A study by Palokangas et al. titled "Measuring the Readiness to Purchase using Methods of Neuromarketing" sought to better understand how marketing tools used in retail stores affect a customer's level of willingness to buy at various stages of the consultative selling process. Instead of the valuation mechanism that assesses the risks associated with the option, the expected subjective preferences and reward may not be the deciding considerations throughout the consultative selling process.

Sola's article, "Neuromarketing: Science and Practice," outlines the technological and general constraints on doing research that have an impact on product development, graphic design, distribution, and general product/service marketing. Entrepreneurs can use the data analysis provided by the discipline of neuromarketing, which is a beneficial addition to marketing research, to create and market products that are in accordance with consumer preferences.

In order to outline the impact of neuromarketing on product presentation at the point of sale in the retail sector, Berger et al. conducted a study on the effect of neuromarketing elements at the point of sale on the consumer's decision-making process in the retail sector. The findings showed that customers with a higher dwelling time buy more and customers also buy more when they taste the products.

When shoppers purchase both items for themselves and a present, they also spend more money.

In their paper, "A Study on Neuromarketing as Emerging Innovative Market Research Strategy: An Interpretive Structural Model Approach," Madan et al. aimed to pinpoint the major factors influencing the acceptance of neuromarketing in Indian society. According to the study, there are many variables that may be managed that affect whether neuromarketing is accepted. To accomplish the dependent variables, entrepreneurs should concentrate on implementing methods that place more of emphasis on driving variables.

A study titled "Neuromarketing: No Brain, No Gain" was published in by Nyoni et al. The study's goal is to shed light on the neurological components of market research, which are generally ignored in conventional marketing research. An examination of the professional difficulties and moral dilemmas in neuromarketing, along with suggestions for the future, is covered in the study's conclusion.

A research investigation titled "Perception of Consumers Regarding Neuromarketing in India with Specific Emphasis on the Kano Model" was conducted by Kumar et al. The core objective of this study was to assess how Indian consumers perceive the concept of neuromarketing. The Kano Questionnaire was employed as the research method, encompassing inquiries related to consumer awareness, consent, and comprehension regarding potential privacy breaches. The study's outcomes unveiled that Indian consumers consider neuromarketing crucial for comprehending consumer behavior and purchase choices while simultaneously enhancing the efficiency of marketing endeavors. When neuromarketing practices are implemented covertly in India, it leads to consumer discord, aligning with their distinct awareness of this methodology.

2.1 Research Gaps

The research purports to understand the consumer's response and their psychology along with the advancements in this area but still fails to come up with better requirements and better tools in order to detect consumer behavior.

The study can be utilized to discover strategies to improve the efficacy of other neuroscientific research techniques. Since the justifications for people's decisions are arbitrary, change over time, and differ from person to person, there needs to be a study connecting this component with pertinent results from a neurological standpoint. Whatever the case, each person differs from the next and has distinct preferences. Through your commercials or marketing techniques, you might not be able to satisfy everyone's needs with your good or service. The end result is that the public desire information, but they only want basic information served easily. It costs a lot of money to create virtual worlds in a lab, and it takes a lot of time to observe how people behave.

2.2 Research Questions

1. Can neuromarketing be one of the advertising strategies in marketing?
2. What is the neuromarketing and purchase decision relationship?
3. Do customer perception influence their purchase decisions?

2.3 Objectives of the Study

- To understand the key concept of neuromarketing and its aids in marketing strategy
- To investigate the impact of neuromarketing strategies on sales.
- To explore the impact of neuromarketing strategies on customer satisfaction
- To evaluate the impact of neuromarketing strategies on long-term customer engagement

2.4 Scope of the Study

Businesses can be a turnaround by differently using advertisement and marketing strategies. With the help of this, they can easily get and attract the consumer's attention. Neuromarketing is one of the strategies that keep track of and map all consumers' attention with regard to their placement and position in figures and graphic text attention. They definitely help read the advertisement to get information. So as the consumers are positively impacted, the sale increases due to an increase in purchase.

3 The Model

In this study, we have used both qualitative and quantitative methods to evaluate our abovementioned objectives.

Neuromarketing as a New Era in Marketing
Monitoring consumer breathing rates, monitoring pupil size, and tracking neural signals When utilized ethically and effectively by marketers, neuromarketing is quite the reverse of what it can initially appear to be—it can customize and enhance the consumer experience. Neuromarketing is the study of how marketing influences people's brain activity.

Neuromarketing is the fusion of marketing and neuroscience for the purpose of forecasting and influencing consumer behavior. Neuromarketing approaches and technologies, according to the US National Library of Medicine, identify advertising elements that elicit favorable and unfavorable responses about particular goods and brands. These methods can be used by businesses to establish marketing strategies, create packaging, and hone and perfect items so that customers get exactly what they want. They come in a variety of forms.

Today's marketing and design professionals use neuromarketing to influence their clients' purchasing decisions and better understand what makes them tick.

3.1 Neuromarketing Aids in Marketing Strategies

The world of neuromarketing applies various strategies in the market to understand the consumer's behavior as well as to add value to their customer satisfaction level. The strategies can be as follows:

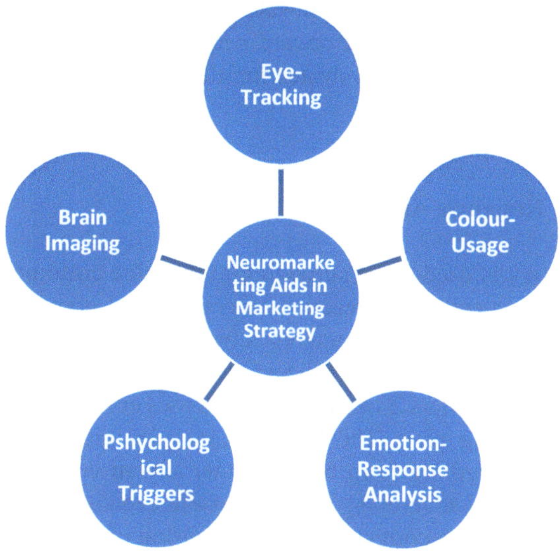

3.1.1 Eye-Tracking

The emotional response of a consumer to marketing and advertising campaigns can be determined through eye-tracking equipment, which analyzes visual focal spots and excitement associated with pupil dilation. The information generated by eye-tracker systems, which can include tools like heat maps and gaze plots, can assist campaigns and brand assets, such as website design, inventive advertising, and product packaging, be better.

GSK, a provider of consumer healthcare, is fully committed to this technology. In the Shopper Science Lab in London, researchers use eye-tracking technology to observe customers as they interact with merchandise during a simulated shopping experience. To assess branding attributes like product description readability and packaging focal points, it evaluates extended fixation (seeing an item for a longer period of time than usual) and repeated fixation (reading or watching an item more than once).

3.1.2 Color Usage

Consumers make subconscious judgments about products "within 90 seconds of initial viewing, and that between 62 percent and 90 percent of that assessment is based on color alone," according to a study by Colorcom.

In his TEDx lecture on neuromarketing, Terry Wu says that as part of its investigation into user engagement with Google ad links, Google tested over 50 colors of blue text to see if it affected clicks. Its later switch to a purple hue of blue for the writing led to an increase in sales of $200 million.

Not a bad outcome for a straightforward neuromarketing strategy. It serves as a reminder for companies creating websites for clients to consider color palettes and the tones utilized in CTA buttons.

3.1.3 Emotion Response Analysis

Webcam technology is used in facial coding to examine customers' facial expressions and gauge their emotional reaction to content and branding. Marketers can adjust their ads to get the intended outcomes by using data that identifies important emotions including anger, contempt, disgust, fear, joy, sadness, and surprise.

Packaging Behemoth Mars is one business that uses emotion response technology as part of its marketing plan. The business analyzes viewer response to video content, including everything from brief Internet advertisements and social media posts to longer-form TV advertising, using its Agile Creative Expertise (ACE). The findings have enhanced the company's comprehension of consumer preferences for content in today's attention-starved culture.

In the world of UX, tools for facial recognition and emotion response analysis are also frequently employed.

3.1.4 Psychological Triggers

Neuromarketing techniques that impact consumer decision-making at an affordable cost include psychological triggers that elicit an emotional reaction. The premise that people would comply in order to be liked by, identified with, or accepted by an influencer or society is one of the most effective ways. Neuromarketing has "demonstrated the relevance of the so-called social impact in social networks: users prefer to copy the activities of others," according to a *Frontiers in Psychology* study. One psychological trigger that is strongly related to social proof is fear of missing out (FOMO).

Showing strong demand for an agency's or brand's goods or services is a straightforward yet effective technique to elicit some of these feelings.

3.1.5 Brain Imaging

The biggest—and most expensive—developments in neuromarketing are in brain imaging technologies. The instant a consumer is exposed to a product or marketing content, their brain activity is scanned. Functional magnetic resonance imaging (fMRI) and electroencephalography (EEG) are the main instruments employed in this study, with functional near-infrared spectroscopy (fNIRS) serving as a more portable and practical alternative.

With the help of this technology, businesses can get the precise, scientific feedback they need on their goods, ideas, and services—feedback that customers can't always successfully express.

According to sales data collected 3 years later, a song's future popularity as indicated by sales data was significantly correlated with "activity in a specific brain area, examined by fMRI while people were listening to music," according to a 2012 Emory University study.

3.2 Importance of Neuromarketing in Today's World

Neuromarketing is a growing field of research that is gaining attention from businesses and marketers alike in cities like Varanasi in Uttar Pradesh, India. Neuromarketing combines neuroscience and marketing to understand consumer perceptions, behavior, and responses to products or services. It seeks to identify how consumers respond emotionally or cognitively to brands and products, as well as analyze their attention spans when it comes to marketing messages.

The aim of neuromarketing is simple: To better understand what triggers the buying decisions of consumers so that marketers can deliver tailored campaigns based on those insights. Neuromarketing focuses on understanding how people make decisions by studying their brain activity—essentially providing insight into what motivates them. Several methods are used in this area such as EEG (electroencephalography), MRI (magnetic resonance imaging), and fMRI (functional magnetic resonance imaging) studies, eye-tracking devices, facial expression analysis, etc., allowing researchers access into the human brain's response system with regard to certain stimuli.

In Varanasi City, there are lots of small-scale industries such as the garment industry that use traditional advertising tools for promotion activities, but now due to technological advancements, they prefer modern promotional tools like neuromarketing for increasing sales volume. In terms of neuromarketing, most appropriate tool for SSI's will be using Eye tracking devices which enable them to measure customer's Attention towards advertisements/promotional displays at shops its very helpful information from viewpoint seller & Buyer both parties get to benefit from it, seller get feedback about his display & buyer get Idea about the product he already has seen so he able take perfect decision whether purchase product or not. Moreover, retailers often use emotions while designing displays or advertising EEG machine help retailers find out an emotional response from customers due visual elements colors, textures, etc. It also helps guess whether customers would remember advertisements after sometime known memory retention.

When it comes to influencing the perceptions of customers in small-scale industries in Varanasi, neuromarketing plays a vital role. It helps businesses capture the attention of customers and keep them engaged for longer periods of time by using visual marketing techniques like animation or product placement. Furthermore, these techniques help create more meaningful associations between products and services, giving the customer a better understanding of what they're buying.

Another key benefit of neuromarketing is its ability to measure responses subconsciously, allowing businesses to gain valuable insights into how their target audience perceive their offerings without having access to verbal feedback from customers directly. Such insights are invaluable for making informed decisions on marketing strategies based on true customer sentiment instead of guesswork or assumptions about what may be effective with consumers in Varanasi's small-scale industry sector.

Not only does neuromarketing enable marketers to understand what message resonates with their target audience, but it also allows them create campaigns tailored specifically toward this audience for maximum impact—something which would otherwise not be possible if traditional marketing methods were relied upon alone. More importantly, it enables brands accurately track exactly where, when, why, and even how customers interact with their offerings over time. This helps formulate future plans based on real-time data rather than relying solely on assumptions.

However, it is important that marketers remember that simply employing sophisticated technology doesn't guarantee success; there are still many more factors involved such as creating compelling content, which grabs shoppers' attention while providing them value at the same time. Furthermore, companies must ensure they use reliable sources while collecting data so they can make unbiased decisions regarding their strategy while adhering strictly to ethical standards when dealing with sensitive information related to consumer behavior.

Overall, neuromarketing provides numerous benefits for companies attempting to influence consumer perception within small-scale industries in Varanasi through visual advertising techniques coupled with analysis backed up by scientific research—something that traditional methods may sometimes fall short of delivering efficiently. It presents an opportunity for companies seeking a competitive edge through innovative means while staying compliant within legal frameworks set forth both nationally and internationally, hence enabling brands to remain successful without compromising quality standards expected from public entities operating at large scales across borders.

Eye-tracking technology has been used extensively in the market research sector for many years, but recently its applications have multiplied across industries, from retailing to space design review to medical product testing to gaming user experience design. Through eye-tracking, users come up with more creative solutions faster than ever before, and even Web development companies use this technology to optimize Web pages by analyzing where users focus their eyes.

Additionally, new technologies like artificial intelligence (AI), computer vision, deep learning, artificial neural networks, big data, predictive analytics, etc. could assist retailers in tracking customer behavior as well as predicting future trends, enabling them to take proactive measures to achieve desired results: cost-effective strategic planning, more effectiveness, creativity, sustainability, etc.

However, it is now understood that judgments, especially those involving purchases, are the result of stimuli and our brain's reaction when it interacts with components to produce a positive response. Marketers utilize consumer neuroscience as a technique to study customer behavior and its effects. However, it is now understood that judgments, especially those involving purchases, are the result of stimuli and our brain's reaction when it interacts with components to produce a positive response. Marketers utilize consumer neuroscience as a technique to study customer behavior and its effects.

According to the *Harvard Business Review*, the study of consumer neuroscience, also referred to as neuromarketing, involves looking into the brain to forecast and possibly even influence consumer behavior and decision-making. Researchers who study neuromarketing keep an eye on brainwaves and chemical shifts to find physiological and neural signs that can help them better understand the motives, preferences, and decisions of consumers. Pricing, product development, and other marketing strategies can be improved with the use of this information. In the middle of the 2000s, consumer neuroscience had a surge in popularity as business school researchers began to show that branding, advertising, and other marketing practices can have measurable effects on the brain. In 2004, Coca-Cola and Pepsi were given to participants in an fMRI scanner by researchers at Emory University.

The scientists detected a recurring brain response when the beverages' identities were hidden. The limbic structures, which are parts of the brain linked to emotions, memories, and unconscious processing, however, displayed more activity when respondents were able to see the brand, indicating that brand knowledge altered how the brain perceived the beverage. Neuromarketing has not been widely adopted despite these findings and the unavoidable rise in overall sales, which emphasizes the concerns regarding the validity of brain scans and neural information synthesis.

Despite the fact that people are skilled at conveying their wishes, neuroscience can aid in our understanding of the subtle influences on our decisions. This information can be used to gauge how customers value an item and to establish how and when a business can use displays and its brand to sway a customer's decision. Sumeru Marketing is pushing this revolutionary notion to the forefront. The right elements are added to each of our ads and displays to maximize impact using heat maps from neuroscience.

3.3 Future of Neuromarketing: The Future of Marketing Research

Even while the past of neuromarketing hasn't quite lived up to the hype, the future is promising. Neuromarketing can be successfully applied in the future by collaborating with virtual reality (VR). The use of VR headsets in the gaming business has already increased, and now Open BCI, a neurotech company, is attempting to embed an EEG device into a VR headset. This technology would allow the company to gather neurological information from the headset user. Even while the company is primarily developing this VR/EEG equipment to learn how to create better games in the future, it has significant implications for marketing research.

What if, for example, you could conduct a marketing research study without having to pay a high rental fee for a fancy venue by just giving participants VR headsets? People can view whatever you want them to while wearing this headgear, and you can keep an eye on their neurological signals as they process the information. Let's go back to the example of the clothing company where you are attempting to gauge consumer acceptance of a new line.

What if the study participant could put on their headset and look at a virtual version of the shop? The participant will then begin to browse the "shop," and you will start to notice brain signals that will tell you how they are reacting to the new line. In other words, depending on their brain signals, you could identify if a person was happy, bored, frustrated, and more! Your company will receive crucial information from this regarding whether or not to introduce the line. The headgear might also be paired with an eye-tracker. Therefore, during each change in brain activity, neuromarketing would be able to tell precisely where the user was looking.

If a participant's brain emits "happy" signals when gazing at a pair of the new line's trousers but "bored" signals when looking at a top, for example, the clothing manufacturer would find out about this information useful. Instead of discarding the entire line, it could be preferable to use a few of the components. Despite the fact that we have not yet arrived, I believe that this technology will be here sooner than we anticipate because it is already under development.

If VR/EEG headgear are made available to the public, one can only conjecture about what the distant future may entail. Participants may not need to wear anything on their heads in the future; instead, they will only need to wear contact lenses with pupil dilation sensors. Depending on whether one is witnessing something favorable or unfavorable, their pupils' sizes alter. This means that in addition to being able to detect when a participant discovers something they like or dislike, these contact lenses may also be able to tell a neuromarketer whether a company should run a commercial or launch a new clothing line. Smart glasses could be used to measure EEG activity. In fact, a team of academics in the USA and Korea are already developing smart glasses with EEG measurement capabilities.

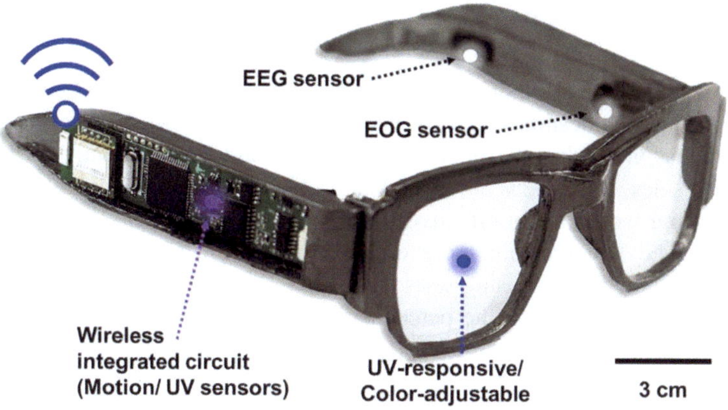

Let's go back to the online store scenario for a second. Imagine if I told you that, in the future, the participant might not even be required to don contacts, a headset, or smart glasses. Instead, what if we could simply implant the virtual shop within their brains? Resembles a science fiction movie? Elon Musk's company, Neralink, is creating neural implants that work as BCIs for telepathy, gaming, memory storage, and replay, among other uses. Even while it will probably take a few years before this technology can be used for marketing research, I do think that neuromarketing will eventually develop in this direction.

3.4 Neuromarketing in SSI in Varanasi

Neuromarketing has become an increasingly popular technique used by small-scale industries in Varanasi to understand consumer behavior, shape perceptions, and influence decisions. By utilizing modern neuroscience technologies, such as functional magnetic resonance imaging (fMRI) and electroencephalography (EEG), neuromarketing can provide businesses with powerful insights into how consumers think and feel when exposed to certain products or services.

When it comes to influencing the perceptions of customers in small-scale industries in Varanasi, neuromarketing plays a vital role. It helps businesses capture the attention of customers and keep them engaged for longer periods of time by using visual marketing techniques like animation or product placement. Furthermore, these techniques help create more meaningful associations between products and services, giving the customer a better understanding of what they're buying.

The other objectives of the research paper are to investigate the impact of neuromarketing strategies on sales, customer satisfaction, and long-term customer engagement in small-scale industries in Varanasi.

We have used descriptive research. Questionnaire method has been used to collect the data, and a fair representative of the population from Varanasi is selected, consisting of 140 respondents.

4 Empirical Results

The demographic analysis of respondents is presented in Table 1 (Graph 1).

On posing the question impact of neuromarketing on sales, 42% of respondents believed that increasing the price of products would be an effective strategy for increasing sales. While 32% of respondents believed that using persuasive language in advertising would be an effective strategy for increasing sales, 26% of respondents believed that creating more effective marketing campaigns would be an effective strategy for increasing sales (Graph 2).

The above figure shows the percentage of respondents who provided different answers to the question "What is customer satisfaction?"

Table 1 Demographic representation of respondents

SI. no	Demographical variables	Frequency	Percentage
1.	*Gender diversification*		
	Male	78	65
	Female	62	35%
	Total	*140*	*100*
2.	*Age distribution*		
	>18	4	3
	19–25	56	40
	26–35	56	40
	36–45	20	14
	>45	4	3
	Total	*140*	*100*
3.	*Educational qualification*		
	Postgraduate	43	31
	Graduate	61	44
	Undergraduate	24	17
	Others	12	9
	Total	*140*	*100*
4.	*Occupation*		
	Non-salaried	30	21
	Salaried	56	40
	Self-employed professional	43	31
	Self-employed non-professional	11	8
	Total	*140*	*100*
5.	*Monthly income*		
	Up to 30,000	46	33
	30,001–60.000	55	39
	60,001–1,00,000	17	12
	Above 1,00,000	22	16
	Total	*140*	*100*

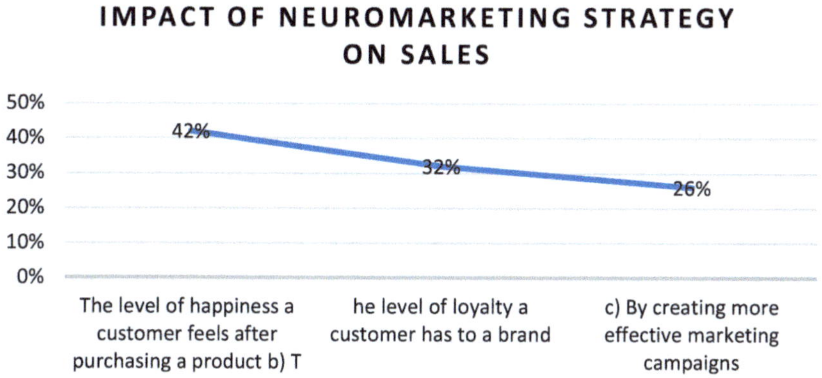

Graph 1 Respondent representation on the impact of neuromarketing on sales

Graph 2 Customer satisfaction as per respondent

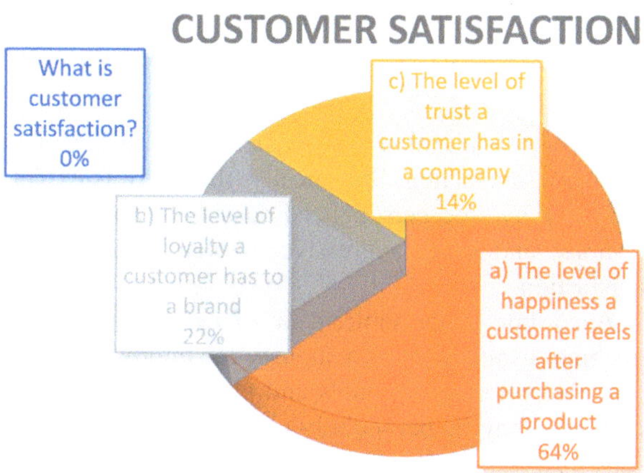

Sixty-four percent of respondents believed that customer satisfaction is the level of happiness a customer feels after purchasing a product. This suggests that the respondents view customer satisfaction primarily as an emotional response to the product or service purchased and that positive emotions like happiness are key indicators of customer satisfaction.

Twenty-two percent of respondents believed that customer satisfaction is the level of loyalty a customer has to a brand. This suggests that the respondents view customer satisfaction as a longer-term relationship between the customer and the brand and that loyalty is a key indicator of satisfaction. On the other hand, 14% of respondents believed that customer satisfaction is the level of trust a customer has in a company. This interpretation suggests that the respondents view customer satisfaction as an indicator of the customer's level of trust in the company providing the product or service and that trust is a key factor in satisfaction (Graph 3).

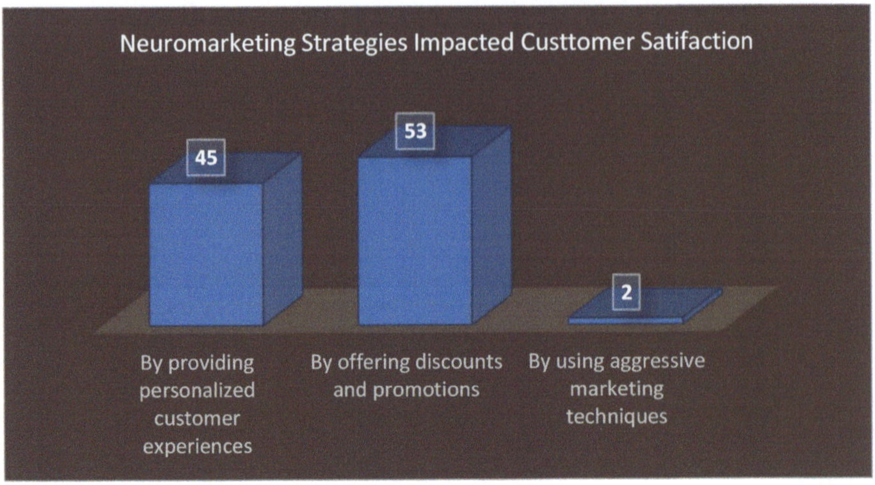

Graph 3 Respondent perception toward neuromarketing strategies

The above question is examining the impact of neuromarketing strategies on customer satisfaction. Specifically, the study is exploring how different strategies, such as providing personalized customer experiences, offering discounts and promotions, and using aggressive marketing techniques, have influenced customer satisfaction.

According to the table, 45% of respondents identified providing personalized customer experiences as a strategy for improving customer satisfaction, while 53% identified offering discounts and promotions. These results suggest that these two strategies may be particularly effective in improving customer satisfaction, potentially as a result of neuromarketing techniques.

Interestingly, only 2% of respondents identified using aggressive marketing techniques as a strategy for improving customer satisfaction. This could suggest that these techniques are less effective in improving satisfaction or that they may even have a negative impact on satisfaction (Graph 4).

The above figure represents the percentage of respondents who identified different factors that contribute to long-term customer engagement. Forty-one percent of respondents believed that the level of repeat purchases from a customer is an important factor in long-term customer engagement. This could be interpreted as an indication that these respondents believe that customers who make repeat purchases over time are more likely to be engaged with a brand and to continue to be customers in the long term. Forty-two percent of respondents believed that the level of customer loyalty to a brand is an important factor in long-term customer engagement. This could be interpreted as an indication that these respondents believe that building customer loyalty is crucial to maintaining long-term customer engagement. This could involve creating a positive brand image, delivering high-quality products or services, and offering excellent customer service, while only 17% of respondents

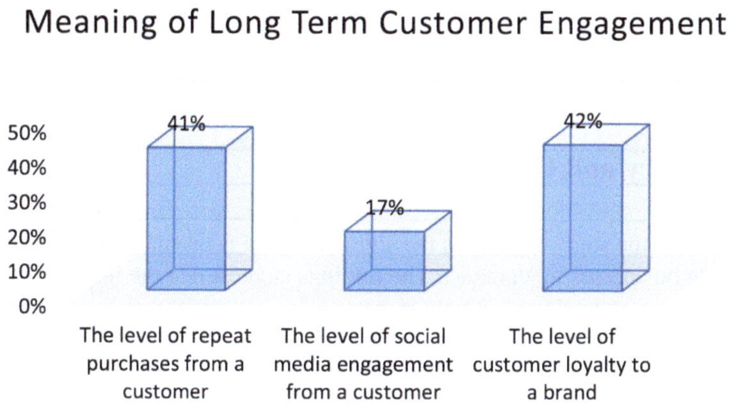

Graph 4 Respondent awareness toward long-term customer engagement

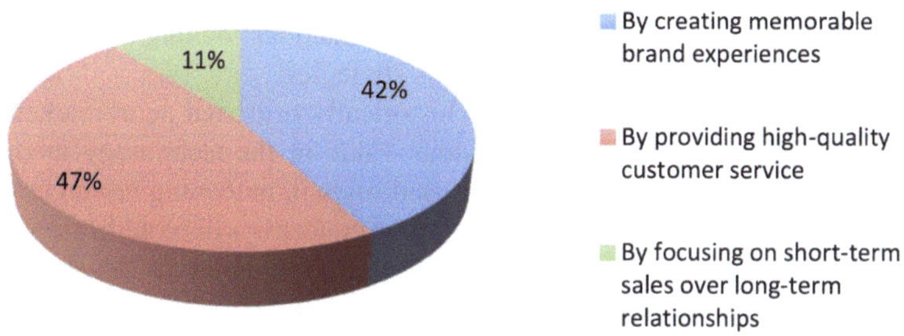

Graph 5 Impact of neuromarketing strategies to long-term customer engagement

identified the level of social media engagement from a customer as an important factor in long-term customer engagement. This suggests that social media engagement may be seen as less important than other factors, such as repeat purchases and customer loyalty (Graph 5).

The figure exhibits the percentage of respondents who identified different ways in which neuromarketing strategies have impacted long-term customer engagement.

Forty-seven percent of respondents identified providing high-quality customer service as a way in which neuromarketing strategies have impacted long-term customer engagement. This suggests that using neuromarketing techniques to improve customer service may be an effective way to build long-term customer engagement, while 42% of respondents identified creating memorable brand experiences as a way in which neuromarketing strategies have impacted long-term customer engagement. This could involve using techniques such as sensory marketing, emotional branding, and storytelling to create a memorable and engaging brand experience for customers.

Only 11% of respondents identified focusing on short-term sales over long-term relationships as a way in which neuromarketing strategies have impacted long-term customer engagement. This suggests that this approach is seen as less effective in building long-term customer engagement.

5 Summary and Conclusions

In conclusion, the study highlights the significance of neuromarketing strategies for small-scale businesses in Varanasi. The findings suggest that the implementation of neuromarketing strategies positively impacts sales, customer satisfaction, and long-term customer engagement. Through the analysis of primary data collected from 140 respondents, it was found that the implementation of neuromarketing strategies positively impacted all three areas of interest, leading to increased sales, improved customer satisfaction, and greater long-term customer engagement.

Ultimately success strategy depends on the correct application of relevant technologies engaging interactive experiences personalizing communication strategies should remain the prime objective for all organizations looking to achieve optimal results through marketing communications strategies.

Though budgetary restrictions have historically restricted neuromarketing, marketing professionals may find tremendous value in the technology in the future. VR/EEG headsets will likely start to be used often in marketing research during the next 5 years, in my opinion. I believe that these concerns will be addressed before the technology covered in this article is widely used, despite the fact that there may be some real concerns regarding privacy, safety, and other issues. Overall, this study provides important insights into the benefits of neuromarketing strategies in enhancing business performance and provides practical recommendations for businesses to effectively implement these strategies in their marketing campaigns.

6 Limitations and Scope for Further Study

While this study provides valuable insights into the impact of neuromarketing strategies on small businesses, it is not without limitations. Firstly, the study was conducted in Varanasi and may not be representative of other regions or contexts. Additionally, the study relied solely on primary data collected from a limited number of respondents, which may not be generalizable to a larger population.

Therefore, future studies could aim to replicate this study in different regions and with a larger sample size to determine if the findings hold true across different contexts. Furthermore, this study focused solely on small-scale businesses, and future research could examine the impact of neuromarketing strategies on larger organizations.

Moreover, while the study identified challenges faced by businesses in implementing neuromarketing strategies, further research could investigate in greater detail the reasons behind these challenges and how they can be overcome.

Additionally, this study only investigated the impact of neuromarketing strategies on sales, customer satisfaction, and long-term customer engagement. Future studies could explore the impact of these strategies on other important business metrics, such as brand loyalty, customer lifetime value, and profitability.

References

1. Singh, M., & Singh, R. K. (2019). Consumer neuroscience: Understanding the neuromarketing concept. *Journal of Marketing and Communication, 15*(2), 8–14.
2. Yadav, S., & Arora, A. (2018). An empirical study on the influence of neuromarketing on consumer behavior. *International Journal of Research in Finance and Marketing, 8*(2), 15–28.
3. Srivastava, M., & Singh, A. K. (2016). A study on neuromarketing: A new dimension of marketing research. *International Journal of Advanced Research in Management and Social Sciences, 5*(2), 24–38.
4. Chakraborty, S., & Das, G. (2018). Neuromarketing: An emerging trend in marketing research. *International Journal of Business and Management Invention, 7*(6), 54–58.
5. Garg, R., & Gupta, M. (2019). Role of neuromarketing in decision making: A conceptual study. *International Journal of Scientific Research and Review, 8*(2), 580–586.
6. Kumar, A. (2019). Influence of neuromarketing on consumer buying behavior: A review. *International Journal of Engineering, Applied and Management Sciences Paradigms, 1*(2), 54–58.
7. Singh, P. K., & Kharub, M. K. (2017). A study on neuromarketing: A new era of marketing. *International Journal of Management Research and Review, 7*(9), 936–942.
8. Tiwari, A., & Khanna, A. (2019). Exploring the impact of neuromarketing on consumer behavior: A conceptual study. *International Journal of Engineering, Applied and Management Sciences Paradigms, 3*(3), 16–21.

A Descriptive Study on Artificial Intelligence and Integrity: Challenges and Prospects

M. Chandrakala and C. H. Raja Kamal

Abstract In the coming decades, machine learning and artificial intelligence (AI) technologies will continue to fundamentally transform civilization. The powerful new technologies that are revolutionizing society will have a big ethical effect on how they both improve and interfere with human life. Because AI is the externalization of human intelligence, it offers us everything that mankind already is—both good and bad—in enhanced form. The stakes are high. We must carefully evaluate how to effect this transition at this critical juncture in history lest we run the risk of fostering the more sinister, rather than the more charitable, sides of our nature.

Keywords Artificial intelligence · Machine learning · Disrupting human lives · AI-powered surveillance · Moral deskilling

1 Introduction

Why are AI ethics suddenly a hot topic? Three factors explain why neural network-based machine learning (ML) is expanding quickly: (1) A substantial growth in the number of data sets, (2) a substantial increase in computing power, and (3) a substantial improvement in machine learning algorithms and a substantial improvement in the human ability to design them [1]. These three inclinations all involve the concentration of power, and tremendous power comes with great responsibility, as the proverb says. For many years, the Institution for Applied Ethics has given serious consideration to the morality of AI. The original version of this article was delivered as performances at academic conferences [2]. Since then, it has developed into an academic paper (see links below) and, most recently, an appearance of "Artificial Intelligence and Ethics: Sixteen Issues," which I have given both domestically and abroad, in that vein.

M. Chandrakala · C. H. Raja Kamal (✉)
Department of Commerce, Kristu Jayanti College (Autonomous), Bangalore, India

J. M. K. P. et al. (eds.), *Business Resilience and Digital Technology in the Post-Pandemic Era*, Contributions to Environmental Sciences & Innovative Business Technology, https://doi.org/10.1007/978-3-031-48075-1_14

2 Technical Safety

Any technology must first show that it works the way it was designed to. Will AI systems perform as expected or not? If and when they collapse, what will happen? How will it fail? If we rely on them, will we be able to survive without them?

For instance, when a semi-autonomous car encountered situations in which it was unable to make a safe judgement, the accident resulted in the deaths of multiple people. However, from a moral perspective, not only is the manufacturer still responsible, but the contract itself may be seen as a dishonest attempt to escape legal liability [3]. Legally, a manufacturer's responsibility may be diminished by creating very specific contracts that place strict limitations on liability.

3 Transparency and Privacy

Transparency: AI systems should strive to be transparent, meaning their inner workings and decision-making processes should be explainable and understandable to human users. Transparency helps build trust in AI systems and allows users to have a better understanding of how the technology operates. Transparent AI systems also enable accountability, as users can identify biases, errors, or unethical behavior and take appropriate actions [4].

3.1 Privacy

Protecting privacy is crucial when using AI. AI systems often require access to vast amounts of data to function effectively, and this can include personal and sensitive information. Organizations must ensure that data collection, storage, and usage adhere to privacy laws and regulations [5]. Data anonymization and encryption techniques can be employed to minimize privacy risks. Additionally, clear consent mechanisms should be in place to ensure individuals have control over the use of their data [6].

3.2 Artificial Intelligence

AI refers to the development of computer systems that can perform tasks that typically require human intelligence. AI systems are built using algorithms and models that can process large amounts of data, learn from patterns, and make decisions or predictions [7]. The integrity of AI systems involves ensuring that they are designed and trained in an ethical manner, free from biases and discrimination. It is important to regularly evaluate AI systems for fairness, robustness, and accuracy to maintain integrity.

3.3 *Integrity*

In the context of AI, integrity refers to the ethical behavior and reliability of the systems. AI should be designed to operate with integrity, meaning it should respect human values, adhere to ethical standards, and avoid harmful or malicious actions [8]. The integrity of AI systems can be upheld through rigorous testing, ongoing monitoring, and the inclusion of ethical considerations during the development process.

4 Beneficial Use and Capacity for Good

Beneficial use and the capacity for good are important considerations when it comes to AI and integrity. While AI technologies have the potential to bring numerous benefits and positive impacts, it is crucial to ensure that they are developed and deployed in an ethical and responsible manner [9]. Here's how AI can be used for good while maintaining integrity:

4.1 *Social Good*

AI can be leveraged to address significant societal challenges. For example, it can be used in healthcare to improve diagnostics, develop personalized treatments, and enhance patient care. AI can also help in areas such as disaster response, environmental conservation, poverty alleviation, and education by providing insights, predictions, and innovative solutions [10].

4.2 *Fairness and Equity*

Ensuring fairness and equity in AI systems is essential. Bias can inadvertently be introduced if AI models are trained on biased data or if the design of the system is not inclusive. Developers and practitioners should strive to eliminate bias, discrimination, and unfairness by carefully selecting and preprocessing training data, using representative datasets, and conducting regular audits to detect and mitigate biases [11].

4.3 Privacy and Data Protection

Respecting privacy and safeguarding data are critical for maintaining integrity. AI systems should be designed to protect user data, minimize data collection to what is necessary, and ensure secure storage and transmission of information. Implementing privacy-enhancing technologies, such as differential privacy or federated learning, can help preserve privacy while still allowing the benefits of AI [12].

4.4 Ethical Decision-Making

AI systems should be guided by ethical principles and adhere to established norms. Developers should consider the potential impact of their systems on individuals, communities, and society as a whole. Ensuring transparency and explain ability in AI algorithms can help uncover and address ethical concerns. Ethical guidelines, standards, and regulatory frameworks should be developed and followed to ensure the responsible use of AI.

4.5 Human-AI Collaboration

AI should be viewed as a tool to augment human capabilities rather than replace them. By focusing on human-AI collaboration, we can leverage AI's strengths while considering human values, context, and judgment. This approach helps maintain integrity by ensuring that decisions made by AI systems are transparent [13].

5 Malicious Use and Capacity for Evil

A perfectly functional technology like a nuclear bomb can wreak great evil when employed for its intended purpose. It is unlikely that artificial intelligence will be exploited improperly unlike human intelligence.

For instance, AI-powered surveillance is already widely used in both acceptable scenarios (like airport security cameras) and potentially undesirable ones (like devices with always-on microphones in our homes) (e.g., products that help authoritarian regimes identify and oppress their citizens). Systems for lethal autonomous weapons (LAWS), known as "killer robots," and AI-assisted computer hacking are two more evil examples that can be given. Other worries include situations like those in the films "2001: A Space Odyssey," "Wargames," and "Terminator," with varied degrees of reality [14].

6 Bias in Data, Training Sets, Etc.

Bias in data and training sets is a critical consideration when it comes to AI and integrity. AI systems learn from data, and if the data used for training is biased, it can result in biased or unfair outcomes. Here's a closer look at the issue and how to address it.

6.1 *Understanding Bias in Data*

Bias in data refers to the presence of systematic and unfair preferences or prejudices within the data itself. It can arise due to various factors, such as societal biases, historical discrimination, or uneven representation in the data collection process. Bias can manifest in different forms, including racial, gender, or socioeconomic biases.

6.2 *Impact of Biased Data on AI Systems*

When AI systems are trained on biased data, they can learn and reinforce the biases present in that data. As a result, the AI models may produce biased predictions, decisions, or recommendations that perpetuate unfairness or discrimination. This can lead to negative consequences, such as reinforcing stereotypes, exacerbating social inequalities, or perpetuating discrimination against certain groups.

6.3 *Addressing Bias in Data and Training Sets*

6.3.1 Diverse and Representative Data

Ensuring that training data is diverse and representative is crucial. It involves collecting data that covers different demographics, backgrounds, and perspectives, reducing the risk of underrepresented groups being overlooked or marginalized.

6.3.2 Data Pre-Processing

Pre-processing techniques can be used to identify and mitigate bias in the data. For example, techniques like data augmentation, oversampling, or under-sampling can help balance the representation of different groups in the dataset.

6.3.3 Bias Detection and Mitigation

Employing methods to detect and mitigate bias during the training process is essential. This can involve measuring and analyzing the impact of the AI system's predictions on different groups and taking steps to address any identified biases, such as adjusting the model or modifying the training process.

6.3.4 Ongoing Monitoring and Evaluation

Regularly monitoring and evaluating AI systems for bias is crucial to maintain integrity. This can involve conducting audits and assessments to identify and address any biases that may have emerged after deployment.

6.3.5 Ethical Considerations and Standards

Integrating ethical considerations and standards into the development and deployment of AI systems are vital. Organizations should have clear guidelines and policies that explicitly address bias in data and training sets. Ethical review boards or external oversight mechanisms can provide guidance and ensure the responsible use of AI.

6.3.6 Transparency and Explain Ability

Transparency in AI systems can help identify and address biases. Providing explanations of how the AI system arrived at its decisions can help uncover underlying biases and enable users to assess the fairness and integrity of the system.

7 Unemployment/Lack of Purpose

The rise of AI poses potential risks to various job categories, including traditionally less susceptible fields like law, medicine, and education. While the specific new jobs that may emerge are uncertain, people are more likely to prefer roles involving human interaction and care provision. The concern for work is interconnected with how people spend their time and what motivates them.

8 Growing Socioeconomic Inequality

The growing socioeconomic inequality related to artificial intelligence (AI) raises concerns about fairness, access, and the overall integrity of AI systems. Here are some key aspects to consider.

8.1 Access and Affordability

AI technologies and their benefits may not be equally accessible to all segments of society. Affordability, availability of resources, and infrastructure can create disparities in access to AI tools, training, and opportunities. Unequal access can further exacerbate existing socioeconomic inequalities, as those who lack access may be left behind in terms of educational, employment, or economic opportunities.

8.2 Bias and Discrimination

AI systems can inadvertently perpetuate and amplify existing socioeconomic inequalities if they are trained on biased data or reflect biased societal norms. For example, if AI algorithms are used in hiring or lending decisions, they may inadvertently discriminate against marginalized or disadvantaged groups, reinforcing existing disparities.

8.3 Job Displacement and re-Skilling

AI and automation have the potential to impact the job market and socioeconomic dynamics. While AI can create new job opportunities, there is also a concern about the potential displacement of certain jobs. Without proper re-skilling and support programs, individuals in affected industries may face challenges in adapting to the changing employment landscape, widening the socioeconomic divide.

8.4 Concentration of Power

AI technologies are often developed and controlled by large tech companies or organizations with significant resources. This concentration of power can lead to monopolistic tendencies, limited competition, and unequal distribution of benefits. It is important to ensure that AI technologies are developed and deployed in ways that promote fair competition, accountability, and inclusive participation.

8.5 Ethical Considerations and Regulation

Addressing the socioeconomic inequality associated with AI requires considering ethical considerations and regulatory frameworks. Ethical guidelines can help guide the development and deployment of AI systems, ensuring fairness, transparency, and inclusivity. Regulatory measures can promote competition, prevent monopolies, protect workers' rights, and address potential biases and discriminatory practices.

8.6 Collaboration and Inclusivity

Addressing socioeconomic inequality requires collaboration among stakeholders from various sectors, including governments, technology companies, civil society organizations, and academia. Engaging diverse perspectives and ensuring inclusive decision-making processes can help identify and mitigate potential biases, ensure equitable access, and promote the integrity of AI systems.

9 Environmental Effects

Due to the significant energy requirements, training machine learning models can be expensive, costing tens of millions of dollars or more. It goes without saying that if this energy is derived from fossil fuels, it will negatively impact the climate and be damaging elsewhere in the hydrocarbon supply chain.

Machine learning may greatly improve the efficiency of electrical distribution and consumption in addition to assisting in the search for answers to problems in biodiversity, environmental research, resource management, etc. One method to employ AI's abilities is to improve energy efficiency, because at its core, AI is an efficiency-oriented technology.

The environment may, in general, benefit from AI, but only to a limited extent.

10 Automating Ethics

Automating the ethics of artificial intelligence (AI) and maintaining integrity are a complex challenge, but there are several approaches and considerations to help address this issue.

10.1 Ethical Guidelines and Frameworks

Developing and implementing clear ethical guidelines and frameworks is crucial for automated systems. These guidelines can include principles such as fairness, transparency, accountability, and non-discrimination. They serve as a foundation for designing AI systems with ethical considerations in mind.

10.2 Incorporating Ethical Decision-Making into Algorithms

Ethical decision-making can be integrated into AI algorithms by encoding ethical rules and principles directly into the system's design. This could involve defining specific rules or constraints that the system must adhere to, ensuring that it operates within ethical boundaries.

10.3 Explainable AI (XAI)

XAI techniques aim to make AI systems more transparent and understandable to humans. By providing explanations for the decisions and recommendations made by AI systems, users can assess the ethical implications and integrity of the system's outputs. This transparency allows for human oversight and intervention when necessary.

10.4 Continuous Monitoring and Auditing

Regular monitoring and auditing of AI systems are vital to ensure their ethical behavior and integrity. This can involve evaluating system outputs, detecting biases or discriminatory patterns, and assessing the system's compliance with ethical guidelines. Monitoring and auditing processes should be ongoing to address emerging ethical issues and maintain system integrity.

11 Moral Deskilling and Debility

If we hand over our ability to make judgments to robots, we will become less adept at doing so. The autopilot can handle every part of flying an aircraft, from takeoff to landing, but airline pilots purposely choose to manually control the aircraft during crucial times (such as takeoff and landing) to preserve their piloting skills.

Given that one of the uses of AI will be to either assist or replace people at making particular types of judgments, we should be aware that humans may become less adept at these skills over time (e.g., spelling, driving, stock-trading, etc.). In its most extreme version, if AI begins to make moral and political decisions for us, humans will become less skilled in these areas.

12 AI Consciousness, Personhood, and "Robot Rights"

AI consciousness is the term used to describe a level of subjective awareness and experience that is comparable to human consciousness. While AI systems can demonstrate sophisticated abilities in observation, learning, and decision-making, it is still unclear if they can truly possess consciousness. It is difficult to say if AI can actually experience consciousness because the nature of consciousness and its relationship to physical and biological systems are still poorly understood.

Personhood is a philosophical and legal idea that bestows certain rights and protections on people. Traditionally, personhood has been associated with human beings. Extending personhood to AI systems raises complex questions about their moral and legal standing. Criteria for personhood in the context of AI, such as cognitive abilities, self-awareness,

13 AGI and Super Intelligence

It will be an artificial general intelligence (AGI) and the only other such intellect to exist on Earth at the human level if or when AI reaches human levels of intelligence, performing all that humans can do as well as the average human can.

If or when AGI surpasses human intelligence, it will develop into a super intelligence, a being that may be significantly smarter and more capable than we are and to which humans have only ever been able to relate through myths, religions, and stories.

It's important to note that AI technology is advancing really quickly. Governments and multinational companies are vying for control of AI's potential. Furthermore, there is no reason why AI development should end at AGI.

14 Dependency on AI

Dependency on AI refers to the increasing reliance on artificial intelligence in various aspects of our lives. While AI can offer numerous benefits, it also raises concerns about integrity. In short, it is important to ensure that the use of AI is guided by ethical principles, transparency, and accountability to maintain the integrity of AI systems and mitigate potential risks associated with overreliance on AI.

15 AI-Powered Addiction

AI-powered addiction interventions and applications raise important considerations related to artificial intelligence and integrity. While AI can offer potential benefits in addressing addiction, ensuring ethical integrity in its use is crucial. Here are some key points to consider.

15.1 Data Privacy and Confidentiality

AI-powered addiction interventions may require access to sensitive personal data, such as medical records or behavioral patterns. It is essential to prioritize data privacy and confidentiality to protect individuals' rights and maintain trust. Implementing robust security measures, encryption protocols, and anonymization techniques can help safeguard personal information.

15.2 Informed Consent

Individuals engaging with AI-powered addiction interventions should provide informed consent, understanding the purpose, risks, and limitations of the AI system. Transparent communication about the data collection, storage, and processing methods is essential. Users should have control over their data and the ability to opt out or withdraw their consent.

15.3 Bias and Fairness

AI models used in addiction interventions should be trained on diverse and representative data to avoid biases that could disproportionately impact certain populations. Bias in AI systems can lead to unequal access to treatment or perpetuate social disparities. Regular monitoring and auditing should be conducted to detect and address any biases that may arise.

15.4 Ethical Treatment and Well-Being

AI interventions should prioritize the well-being and best interests of individuals struggling with addiction. The interventions should be designed to provide accurate, evidence-based information, resources, and support. The AI system should be aligned with established ethical guidelines and professional standards in addiction treatment.

15.5 Human Involvement and Oversight

While AI can support addiction interventions, it should not replace human involvement and oversight. Human professionals, such as healthcare providers and counselors, play a vital role in understanding the nuances of addiction, tailoring interventions to individual needs, and providing empathetic support. AI should be seen as a tool to augment human capabilities rather than replace them.

16 Effects on the Human Spirit

The ways in which people view themselves, interact with one another, and live their lives will all be impacted by the aforementioned research areas of interest. A more existential query, though, is also present. Are we rendering ourselves inferior creatures to our own creations if the meaning does identity of mankind have anything to do with our intelligence, if we externalize our intelligence and increase it beyond human intelligence (e.g., as many great Greek philosophers believed)? Artificial intelligence raises a deeper question that goes to the heart of mankind and touches on subjects more typically associated with philosophy, spirituality, and religion. If or when our own creations surpass us in everything we do, what will become of the human spirit? Can human life be lost?

17 Conclusion

In short, maintaining integrity in artificial intelligence (AI) involves addressing ethical considerations, promoting transparency and fairness, and mitigating bias. It requires the development and use of AI systems that align with societal values, respect privacy, and ensure accountability. By upholding integrity in AI, we can strive for responsible and trustworthy AI applications that benefit individuals and society as a whole.

References

1. Green, B. P. (2017, November 21). *Artificial intelligence and ethics: Ten areas of interest.* Markkula Center for Applied Ethics website.
2. Green, B. P. (2017). Some ethical and theological reflections on artificial intelligence. Pacific Coast Theological Society (PCTS) meeting, Graduate Theological Union, Berkeley, 3–4 November, 2017.
3. Green, B. P. (2018, March 21). AI and ethics. Guest lecture in PACS003: What is an Ethical Life? University of the Pacific, Stockton.
4. Green, B. P. (2018, August 24). Ethical reflections on artificial intelligence. *Scientia et Fides, 6*(2), 9–31.
5. Green, B. P. (2019, March 15). *Artificial intelligence, decision-making, and moral deskilling.* Markkula Center for Applied Ethics website.
6. Green, B. P. (2019–2020). *Artificial intelligence and ethics: Sixteen issues.* Santa Clara University.
7. Holt-Lunstad, J., Smith, T. B., Baker, M., Harris, T., & Stephenson, D. (2015). Loneliness and social isolation as risk factors for mortality: A meta-analytic review. *Perspectives on Psychological Science, 10*(2), 227–237.
8. Mannion, L. (2018, January 17). *Britain appoints minister for loneliness amid growing isolation.* Reuters.
9. Markkula Center for Applied Ethics Staff, "AI: Ethical Challenges and a Fast Approaching Future: A panel discussion on artificial intelligence," with Maya Ackerman, Sanjiv Das, Brian Green, and Irina Raicu, Santa Clara University, California, October 24, 2017, posted to the All About Ethics Blog, October 31, 2017.
10. Originally paraphrased in Stan Lee and Steve Ditko, "Spider-Man," Amazing Fantasy vol. 1, #15 (August 1962), exact phrase from Uncle Ben in J. Michael Straczynski, Amazing Spider-Man vol. 2, #38 (February 2002).
11. Scola, L. (2020, May 26). *AI and the ethics of energy efficiency.* Markkula Center for Applied Ethics website.
12. Sylvester, B. (2019, June 26). *Fact check: Did Aristotle say, 'we are what we repeatedly do'?* Check Your Fact website.
13. The Partnership on AI to Benefit People and Society, Inaugural Meeting, Berlin, Germany, October 23–24, 2017.
14. Vallor, S. (2015). Moral deskilling and upskilling in a new machine age: Reflections on the ambiguous future of character. *Philosophy of Technology, 28*, 107–124.

Working Capital Management's Effect on Banks Profitability: Public Sector Banks of India

Ibha Rani and Arti Singh

Abstract The objective of this chapter is to analyze the working capital management's effect on public sector bank profitability. In an economy such as India, the role of public sector banks is very crucial. "Return on assets and return on equity" are representatives of bank profitability, while "current ratio, quick ratio, cash deposit ratio, and credit deposit ratio" have been considered for working capital representative. Secondary data have been employed for the study for 5 years from 2018 to 2022. To check the relationship between working capital factors and bank profitability, correlation and regression analysis has been used. As per analysis, working capital has a significant impact on a bank's profitability. The most influential factor that impacts profitability is cash deposit ratio.

Keywords Profitability · Working capital · Current ratio · Quick ratio

1 Introduction

Working capital management is a key component to maximizing shareholder wealth; at the same time, maintaining liquidity is also an important decision for a bank. It is essential for a sound financial foundation that helps businesses cover their short-term financial obligations and also boost their earnings. The optimum level of working capital helps maximize a firm's value [1]. The optimum level of working capital is the level at which there is a balance between profitability and risk and strikes a trade-off between the risk and return.

I. Rani (✉) · A. Singh
Department of Commerce, Kristu Jayanti College (Autonomous), Bangalore, India

J. M. K. P. et al. (eds.), *Business Resilience and Digital Technology in the Post-Pandemic Era*, Contributions to Environmental Sciences & Innovative Business Technology, https://doi.org/10.1007/978-3-031-48075-1_15

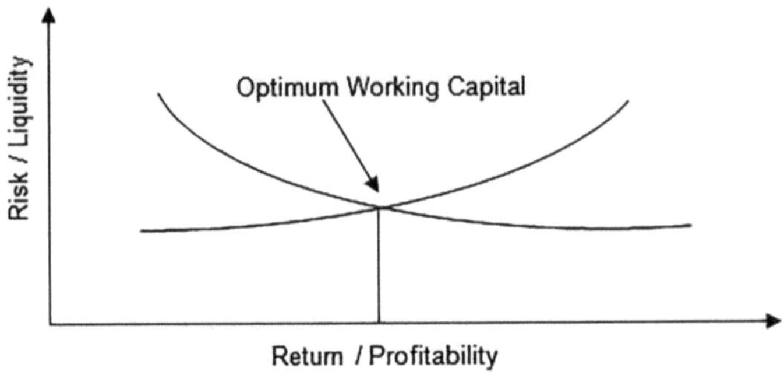

In any firm, including the bank, the financial manager makes three important decisions, finance, investment, and dividend decisions, and these decisions are the pillar of a company. Out of this, finance is the most critical, and it includes both long-term and short-term finance. "Working capital is a short-term financial decision" that gauges the health of a financial institution and is technically calculated as the difference between current assets and current liabilities. A bank's working capital is operating cash on hand to operate the bank on a daily basis.

The banking system plays a crucial role in the modern economic world. An effective banking sector is essential for a nation's development, and India is no exception. It may act as a catalyst for rapid growth in many different economic areas. In India, the Banking sector comprises of the public sector, private sector, and foreign banks. But here, public sector banks have been the backbone of the Indian financial system since the nationalization of the State Bank of India in 1955, followed by more banks in 1969 and 1980. Public sector banks were originated to fulfill social objectives and focus on "banking the unbanked." As per the report, "Basic Statistical Return on Credit by Scheduled Commercial Banks (SCBs) in India – March 2022" released by RBI, public sector banks' share in total credit by the country's commercial bank stood at 54.8%. According to the domestic rating agency, public sector banks contribute 20% to the national income and account for nearly 40% of wages. In the Indian banking industry, we cannot neglect the presence of public sector banks.

In this chapter, we will examine the impact of working capital on bank profitability among all 12 public sector banks in India in 5 years (2018–2022). This study stresses the link between a bank's profitability and its working capital components. Here profitability of banks is considered in the terms of "Return on Assets (ROA) and Return on Equity (ROE)" and components of working capital as "Current Ratio, Quick ratio, Cash ratio, and Credit Deposit ratio."

This article is set up as follows: Sect. 2 outlines the literature review followed by Sect. 3 with the research methodology. Section 4 discusses the impact of working capital and its profitability, and the chapter ends with the conclusion of the study.

2 Literature Survey

Various research papers examined the link between working capital with the performance of banks in different countries which is evident from the UK, Nigeria, Nepal, Pakistan, and other Asian countries. Most of the papers reflect the positive relationship between performance and working capital. Only lesser studies have reflected the negative relationship between working capital and bank profitability. "Working capital management is considered as an important concept in financial management" [2].

Abor [3] investigated the impact of capital structure on the financial firms listed in Ghana's financial market in terms of profitability. Analysis indicates that the "ratio of short-term debts to total assets, and the ratio of total debts to total assets with return on equity" are positively and significantly correlated, while the ratio of "long-term debt to assets and return on equity" have a negative relationship.

Afza and Nasir [4] have determined the positive and significant relationship between profitability and working capital while one of the components that is cash conversion cycle reflects a negative association.

Daniel and Ambrose [5] have assessed the working capital on companies' profitability of manufacturing firms listed on the Nairobi Stock exchange, Kenya. As per the researcher, the firms have invested a large amount of funds in working capital.

Yahaya and Bala [6] studied the Deposit Money Banks in Nigeria for the time frame of 6 years from 2007 to 2013 in terms of performance and their working capital management. They found that the "Cash Ratio" was inversely correlated with "ROA," while "current ratio and quick ratio" were positively correlated to return on assets.

Pravallika and Rao [7] analyzed the performance of the paint industry with the working capital management. As per findings, return on assets and inventories are positively correlated, while ROA are negatively correlated to average receivable days as well as average collection period.

Olaoye et al. [8] examined firms' profitability with working capital, which are listed in the Nigerian Stock Exchange for the period of 2008–2017. The study found that the "cash collection period and cash payment period show a negative impact on ROA."

Senan et al. [9] investigated the 98 Indian banks' performance in view of working capital management from 2008 to 2018. The performance of banks is measured in terms of return on assets, and the most influencer factors of performance are "net profit margin, profit after tax, monetary policy, and working capital cycle."

AlSharif [10] studied the relationship between Islamic bank profitability with "return on deposits, equity multiplier, liquidity ratio, and leverage ratio" for the study period of 2005–2020. As per the statistical tool analysis, there is a statistically significant positive effect at a 5% significance level of return on deposits and equity multiplier on profitability under control variables. Other researchers who have found positive relationships are Pham et al. [11] and Senan et al. [9].

A few studies also witnessed a negative correlation between "working capital management and firms' performance," which include Gill et al. [12], Umer [13], Hailu and Venkateswarlu [14], and Anton [15], while very few studies also suggested no correlation between working capital and profitability, which include Nguyen et al. [16].

2.1 *Objectives of the Study*

The main objective of this chapter is to assess the relationship between working capital and the performance of 12 public sector banks for the period of 5 years 2018–2022.

Hypothesis

H1: Working capital has a significant impact on bank profitability.

H1a: There is a significant relationship between working capital components (current ratio, quick ratio, cash ratio, and credit deposit ratio) and bank profitability (return on assets (ROA).

H1b: There is a significant relationship between working capital components (current ratio, quick ratio, cash ratio, and credit deposit ratio) and bank profitability (return on equity).

3 Research Methodology

3.1 *Sample Banks and Data Collection*

In this chapter, we have assessed the impact of working capital management on the performance of all 12 public sector banks for the period of 5 years from 2018 to 2022. The following banks have been considered for the study (Table 1).

This study is based on secondary data that has been collected from the RBI site, which is the most authenticated source in terms of banking. Table 1 represents the list of the sample banks that have been considered for this study.

Table 1 List of sample banks

S.N	Sample banks	S.N	Sample banks
1	State Bank of India	7	Indian Bank
2	Bank of Baroda	8	Indian overseas Bank
3	Bank of India	9	Punjab and Sind bank
4	Bank of Maharashtra	10	Punjab national bank
5	Canara Bank	11	UCO Bank
6	Central Bank of India	12	Union bank of India

3.2 Measurement of Variables

In this study, to analyze the impact of working capital on bank profitability, for profitability, we have used the return on assets (ROA) and return on equity (ROE), which is the dependent variable for the study. To represent the working capital, current ratio, quick ratio, cash ratio, and cash deposit ratio are used as independent variables (Fig. 1 and Table 2).

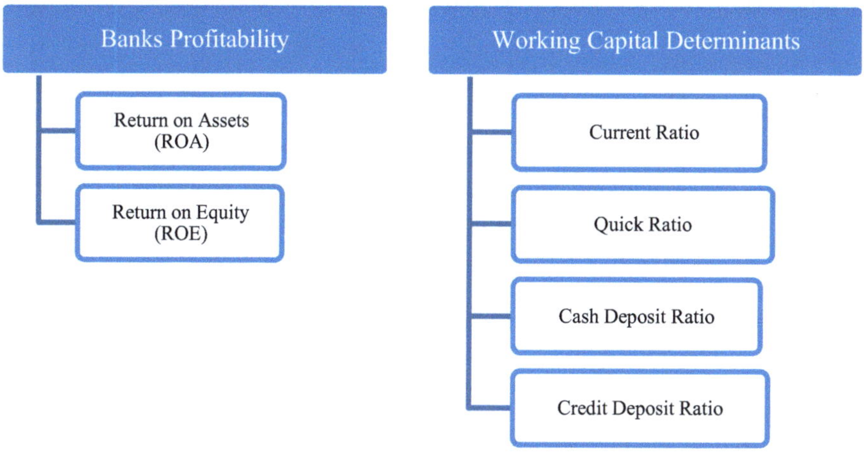

Fig. 1 Bank profitability and working capital Determinants

Table 2 Summary of variables and their calculation

Variables	Identifier	Meanings
Dependent variable	Return on assets (ROA)	It is a ratio that is calculated by banks' net income to the average of total assets
	Return on equity (ROE)	It is a ratio that is calculated by banks' net income to the average shareholders' equity
Independent variables	Current ratio (CR)	It is a ratio that is calculated by current assets to the current liability
	Quick ratio (QR)	It is a ratio that is calculated by current assets subtracted from inventory to the current liability
	Cash deposit ratio (CSDR)	It is the ratio of cash in hand and balances with RBI to the total deposits (demand + time deposits)
	Credit deposit ratio (CRDR)	It is the ratio of total bank credit to the aggregate deposits (demand + time deposits)

3.3 Model Specification

The study examines the relationship between working capital components and bank profitability in India over a 5-year period for 12 public sector banks.

$$\text{Profitability (ROA/ROE)} = \beta0 + \beta1\ (CR) + \beta2\ (QR) + \beta3\ (CSDR) + \beta4(CRDR) + \mu$$

Here $\beta0$ is constant or intercept, while $\beta1$, $\beta2$, $\beta3$, and $\beta4$ are the parameters of working capital, and μ is the residual error.

4 Data Analysis

The sample banks' working capital and its impact on bank profitability have been analyzed through SPSS, and output information is analyzed as follows.

4.1 Descriptive Statistics

In the below table, the minimum, maximum, mean, and standard deviation of data for the dependent and independent variables are represented (Table 3).

As per the Table 3, the average return on assets of sample banks is only 0.638, which indicates that banks have to utilize their assets efficiently to improve their ROA. The average ROE of the sample bank is 6.90, which indicates that banks have efficiently utilized their investors' money to generate profit. While the average values of the independent variable components (QR, CR, CSDR, and CRDR) are recorded as 23.16, 0.727, 5.71, and 63.95, respectively.

Table 3 Descriptive statistics of the variables

Variables	N	Minimum	Maximum	Mean	Std. deviation
ROA	60	−3.01	0.85	0.638	0.943
ROE	60	−61.01	11.86	6.9092	16.718
QR	60	2.76	40.77	23.1653	9.632
CR	60	0.03	0.32	0.0727	0.0478
CSDR	60	1.42	12.21	5.7138	2.20
CRDR	60	47.45	76.04	63.9527	7.771

4.2 Correlation Matrix

The relationship between dependent and independent variables, as well as the relationship between independent variables themselves, is shown in the table below. The relationship is estimated with the help of Pearson's correlation of two-tailed significance. As per analysis, the return on assets which indicates a bank's profitability is positively correlated with the Cash deposit ratio and credit deposit ratio while it is negatively related with the current ratio and quick ratio. Similarly, return on equity also reflects the same pattern as return on assets. The most significant relationship exists between bank profitability and credit deposit ratio. All the independent variables reflect a negative correlation, which means no existence of multicollinearity. But to check the multicollinearity, variance inflation factor (VIF) is conducted, and the value of VIF is less than 5, which represents no problem of multicollinearity (Table 4).

Table 4 Correlations analysis

Variables		ROA	ROE	QR	CR	CSDR	CRDR
ROA	Pearson correlation	1					
	Sig. (2-tailed)						
	N	60					
ROE	Pearson correlation	0.993**	1				
	Sig. (2-tailed)	0.000					
QR	Pearson correlation	−0.064	−0.066	1			
	Sig. (2-tailed)	0.628	0.616				
CR	Pearson correlation	−0.415**	−0.431**	−0.481**	1		
	Sig. (2-tailed)	0.001	0.001	0.000			
CSDR	Pearson correlation	0.016	0.013	0.066	−0.036	1	
	Sig. (2-tailed)	0.903	0.920	0.617	0.784		
CRDR	Pearson correlation	0.274*	0.259*	0.077	−0.254	−0.403**	1
	Sig. (2-tailed)	0.034	0.046	0.557	0.050	0.001	
	Multicollinearity test						
	Variance inflation factors (VIF)			2.3	1.18	1.76	2.12

**Correlation is significant at the 0.01 level (2-tailed)

*Correlation is significant at the 0.05 level (2-tailed)

4.3 Regression Analysis

Here regression analysis has been conducted separately with dependent variables (return on assets) and independent variables as current ratio, quick ratio, cash deposit ratio, and credit deposit ratio (Tables 5 and 6).

The above table reflects the value of R2 as 0.1211, which is a coefficient of determination, and it indicates the total variation in the dependent variable due to the combined effect of independent variables. As per the analysis, 12.11% of the variation in return on assets and 9.21% of the variation in return on equity is accounted due to changes in independent variables as current ratio, quick ratio, cash deposit ratio, and credit deposit ratio.

Table 5 Model estimation as return on assets (ROA) as dependent variables

Model		Unstandardized coefficients		Standardized coefficients		
		B	Std. Error	Beta	t	Sig.
1	(constant)	−0.730	1.297		−0.563	0.576
	QR	−0.033	0.013	−0.338	−2.611	0.012
	CR	−10.276	2.647	−0.521	−3.882	0.000
	CSDR	0.044	0.054	0.104	0.829	0.411
	CRDR	0.025	0.016	0.209	1.618	0.111
	R^2					0.1211
	Adjusted R^2					0.1090
	F-value					5.8
	F-significance					0.001

Dependent variable: ROA

Y1 (ROA)= −0.73 + (−0.033) QR + (−10.276) CR + 0.044 CSDR + 0.025 CRDR-------i

Table 6 Model estimation as return on equity (ROE) as dependent variables

Coefficients[a]						
Model		Unstandardized coefficients		Standardized coefficients		
		B	Std. error	Beta	t	Sig.
2	(constant)	−7.735	22.784		−0.339	0.736
	QR	−0.610	0.222	−0.351	−2.741	0.008
	CR	−192.539	46.486	−0.551	−4.142	0.000
	CSDR	0.682	0.941	0.090	0.724	0.472
	CRDR	0.392	0.276	0.182	1.420	0.161
	R^2					0.0921
	Adjusted R^2					0.08516
	F-value					6.148
	F-significance					0.000

[a]Dependent variable: ROE

Y2 (ROE) = −7.735 + (−0.610) QR + (−192.539) CR + 0.682 CSDR + 0.392 CRDR----ii

According to the analysis, both dependent variables (ROA and ROE) are negatively and significantly related to the current ratio and quick ratio and positively related with the credit deposit and cash deposit ratio.

5 Conclusions

In this chapter, we assessed the impact of working capital management on public sector bank profitability for the duration of 5 years. Here, bank profitability is measured in terms of their return on assets and return on equity, and current ratio, quick ratio, cash deposit ratio, and credit deposit ratios are variables used for working capital management. As per correlation analysis, there is a negative correlation between bank profitability and current and quick ratio, while a positive relationship exists between bank profitability and cash deposit and credit deposit ratio.

Scope for Further Study

The current study assesses the impact of working capital management on public sector bank profitability over a 5-year period, focusing on return on assets and return on equity as measures of profitability and current ratio, quick ratio, cash deposit ratio, and credit deposit ratios as working capital management variables. Further research could extend the study duration to analyze long-term trends in the relationship between working capital management and bank profitability. Additionally, a comparative analysis across different types of banks, such as private sector or foreign banks, could provide insights into the differential impact of working capital management on profitability. Qualitative research methods, such as interviews or case studies, could be incorporated to gain a deeper understanding of the underlying reasons behind the observed correlations. Furthermore, exploring the causal relationships between working capital management variables and profitability through advanced econometric techniques would enhance the study's findings.

References

1. Deloof, M. (2003). Does working capital management affect profitability of Belgian firms? *Journal of Business Finance & Accounting, 30*(3–4), 573–588.
2. Joshi, L. K., & Ghosh, S. (2012). Working capital management of Cipla limited: An empirical study. *International Journal of Marketing, Financial Services & Management Research, 1*(8), 170–186.
3. Abor, J. J. (2005). The effect of capital structure on profitability: An empirical analysis of listed firms in Ghana. *Journal of Risk Finance, 6*(5), 438–445. https://doi.org/10.1108/15265940510633505
4. Raheman, A., & Nasir, M. (2007). Working capital management and profitability–case of Pakistani firms. *International Review of Business Research Papers, 3*(1), 279–300.

5. Makori, D. M., & Jajongo, A. (2013). Working capital management and firm profitability: Empirical evidence from manufacturing and construction firms listed on Nairobi Securities Exchange, Kenya. *International Journal of Accounting and Taxation., 1*(1), 1–14.
6. Yahaya, A., & Bala, H. (2015). Working capital management and financial performance of deposit money banks in Nigeria. *Research Journal of Finance and Accounting, 6*(16), 57.
7. Pravallika, M. V., & Rao. (2018). Critical study on impact of working capital management on profitability of manufacturing industry in India (a study on paint industry). *IOSR Journal of Business and Management, 20*(2), 37–47.
8. Olaoye, F. O., Adekanbi, J. A., & Oluwadare, O. E. (2019). Working capital management and firms' profitability: Evidence from quoted firms on the Nigerian stock exchange. *Intelligent Information Management, 11*(3), 43–60.
9. Senan, N. A. M., Anagreh, S., Al-Dalaien, B. O. A., Almugari, F., Khaled, A. S., & Al-Homaidi, E. A. (2021). Working capital management and banks' performance: Evidence from India. *Journal of Asian Finance Economics and Business, 8*(6), 747. https://doi.org/10.13106/jafeb.2021.vol8.no6.0747
10. AlSharif, B. M. (2021). Working capital management and profitability in Islamic banks: An empirical study. *Indian Journal of Economics and Business, 20*(2), 1–19.
11. Pham, K. X., Nguyen, Q. N., & Nguyen, C. V. (2020). Effect of working capital management on the profitability of steel companies on Vietnam stock exchanges. *The Journal of Asian Finance, Economics and Business, 7*(10), 741–750.
12. Gill, A., Biger, N., & Mathur, N. (2010). The relationship between working capital management and profitability: Evidence from the United States. *Business and Economics Journal, 10*(1), 1–9.
13. Umer, M. (2014). *Determinants of capital structure, a study of manufacturing sector in Pakistan* (pp. 1–38). Superior University Lahore.
14. Hailu, A. Y., & Venkateswarlu, P. (2016). Effect of working capital management on firms profitability evidence from manufacturing companies in eastern, Ethiopia. *IJAR, 2*(1), 643–647.
15. Anton, S. G., & Afloarei Nucu, A. E. (2021). The impact of working capital management on firm profitability: Empirical evidence from the Polish listed firms. *Journal of Risk & Financial Management, 14*(1).
16. Nguyen, A. H., Pham, H. T., & Nguyen, H. T. (2020). Impact of working capital management on firm's profitability: Empirical evidence from Vietnam. *Journal of Asian Finance, Economics and Business, 7*(3), 115–125.

A Study on Online and Store Purchasing Behaviour of Consumers

J. Chandrakhanthan, Surjit Singha, and M. Krishnamoorthi

Abstract The aim of this chapter is to offer an overview of online and store spending decision by matching traditional and digitalized choice of consumers according to their needs and satisfaction and also their preference. This study investigates the consumer's motivation that influences online shopping behaviour and store purchase behaviour. The expected outcome of the research study is to know the relationship between customer's opinion and choices of online and store buying behaviour. Due to busy lives and stressful schedules of majority of people in the twenty-first century, online shopping has taken on a significant role. Online shopping became the most convenient and practical option for them in this situation. However, there are still some who prefer the other mode of shopping based on variety of products, according to trust and quality provided by the vendors.

Keywords Consumer buying behaviour · Store purchasing · Online purchasing

1 Introduction

The fast development of the Web and the approach of Internet business have altered the manner in which customers shop and go with buying choices. With the snap of a button, customers can investigate a huge range of items and administrations, look at costs, read surveys, and make buys from the solace of their own homes. This shift towards Web-based shopping has tested customary physical retailers and provoked analysts to dive further into understanding the elements of shopper conduct in both on-the-Web and disconnected retail conditions.

The investigation of on-the-Web and store buying conduct of shoppers is critical in the present computerized age. It looks to reveal the basic factors that impact customers' dynamic cycles while picking among on-the-Web and disconnected channels, as well as investigating the developing pattern of Omni-channel retailing,

J. Chandrakhanthan (✉) · S. Singha · M. Krishnamoorthi
Kristu Jayanti College (Autonomous), Bangalore, India
e-mail: surjitsingha@kristujayanti.com; krishnamoorthi@kristujayanti.com

J. M. K. P. et al. (eds.), *Business Resilience and Digital Technology in the Post-Pandemic Era*, Contributions to Environmental Sciences & Innovative Business Technology, https://doi.org/10.1007/978-3-031-48075-1_16

where shoppers flawlessly explore between different buying channels. By acquiring bits of knowledge into purchaser conduct, retailers and advertisers can adjust their systems to all the more likely take special care of buyers' inclinations and improve their general shopping experience.

Understanding the inspirations and inclinations that drive shoppers to pick among on-the-Web and store channels is significant for organizations expecting to flourish in an undeniably cutthroat market. Factors like comfort, openness, trust, security, social impact, unmistakable encounters, and social cooperations all assume critical parts in forming customer conduct. By inspecting these variables, scientists can give important bits of knowledge that empower retailers to foster compelling advertising procedures, upgrade their on-the-Web and disconnected presence, and make consistent and connecting with shopping encounters for their clients.

Besides, as innovation keeps on progressing, recent fads and advancements arise that further shape shopper conduct. The ascent of versatile shopping, the effect of expanded reality, and the developing impact of social trade are only a couple of instances of the consistently developing scene in which purchasers explore. Thus, it is fundamental for specialists to constantly explore and dissect these arising patterns to remain at the very front of purchaser conduct research.

Considering these contemplations, this study plans to lead an exhaustive survey of the current writing on on-the-Web and store buying conduct of purchasers. By blending and examining an extensive variety of exploration discoveries, this study looks to give a top-to-bottom comprehension of the key elements driving customer conduct in both on-the-Web and disconnected channels. Furthermore, it will distinguish advancing patterns and future headings for research, empowering organizations to settle on informed choices and adjust their procedures to meet the developing requirements and inclinations of shoppers.

Generally speaking, on-the-Web and store buying conduct of purchasers holds extraordinary importance in the field of showcasing and retail. By revealing insight into the variables impacting buyers' dynamic cycles, this examination adds to a more profound comprehension of shopper conduct and gives significant bits of knowledge to organizations to improve their commitment with customers in an undeniably mind boggling and interconnected retail climate.

2 Literature Survey

Daroch et al. [1] investigated consumer behaviour toward online shopping and also examined various factors limiting consumers for online shopping behaviour. Their study found the difficulties that customers encounter when shopping online.

Amit et al. [2] revealed that due to busy lives and stressful schedules of majority of the people in the twenty-first century, online shopping has taken on a significant role. Online shopping became the most convenient and practical option for them in this situation. The Internet has altered how consumers shop and quickly expanded to a worldwide perspective. Business-to-consumer online shopping is the practise of purchasing goods and services from an online store in a manner that is physically similar to doing so. The classical model behaviour assumption underlies the current work. The behaviour and attitudes of online shoppers in Aizawl are investigated in this essay.

Smith et al. [3] provide a comprehensive overview of the factors influencing consumer purchasing behaviour in both online and offline retail settings. The review identifies convenience, trust, price, social influence, and personalization as key determinants of online purchasing behaviour. In contrast, tangibility, sensory experience, immediate gratification, social interaction, and shopping as an experience are highlighted as important factors influencing store purchasing behaviour.

Johnson and Lee [4] explore the interplay between online and offline shopping behaviour. The review examines studies investigating the effects of online research on offline purchases and vice versa. It also explores the concept of "showrooming" and its impact on consumer behaviour. The review concludes that online and offline channels are interconnected and highlights the need for businesses to adopt an omnichannel approach to meet consumers' expectations.

Chen et al. [5] focus on the role of trust in influencing consumer purchasing behaviour in online and offline contexts. The review examines various trust-building mechanisms employed by businesses, such as secure payment options, customer reviews, and reputation management. It also discusses the impact of trust on purchase intentions, customer loyalty, and satisfaction. The review emphasizes the importance of establishing trust to enhance consumer confidence in both online and store environments.

Wang et al. [6] analyse a wide range of studies to identify factors influencing consumer purchase decision-making in online and store environments. The review highlights convenience, product variety, price, promotions, social influence, and personalization as significant factors affecting online purchasing behaviour. For store purchasing behaviour, tangibility, immediate gratification, social interaction, and shopping experience emerge as key determinants. The review provides a comprehensive understanding of the similarities and differences between online and store purchase decision-making processes.

3 Research Methodology

This study was descriptive in nature.

3.1 Research Objectives

1. To study the online and store purchasing behaviour of the consumers.
2. To identify the factors influenced to buy from online stores.
3. To study the frequency of purchasing through online and store.

3.2 Sample Size

In this study, 113 respondents participated in the survey from Coimbatore.

3.3 Data Collection

Data were collected through structured questionnaires and hypotheses have been created and analysed.

3.4 Hypothesis Developed

Research hypotheses were developed based on the study objectives. The following hypotheses were framed to find out the respondent's acceptance towards audio books:

- H1: Gender has a significant influence on respondents' intention to purchase online.
- H2: Gender does not have a significant influence on respondents' intention to purchase online.

4 Data Analysis

4.1 Demographic Profile

S. no.	Demographic profile	No. of respondents	Percentage
Age group	Below 25	91	80.5
	26–35	16	14.2
	Above 35	6	5.3
Gender	Male	53	46.9
	Female	60	53.1
Education	School level	21	18.5
	UG	43	38
	PG	30	26.5
	Others	19	16.8
Occupation	Government employee	19	16.8
	Private employee	43	38
	Self-employed	30	26.5
	Others	21	18.5

It is observed that 46.9% of the respondents are males and 53% of the respondents are females. 80.5% of the respondents are in the age group of below 25, and 14% of the respondents are in the age group of 26–35 years. About 38% of the respondents are found to be undergrads, while 26.5% belong to post-grad group, and 18.5% have attained school education. Regarding the occupation of the respondents, it is found that 16.8% are government employees, while 38% are private employees, and 26.5 are self-employed.

4.2 *Frequency of Purchasing*

	Online		Store	
Frequency	NoR	Percentage	NoR	Percentage
Often	6	5.3	30	26.5
Weekly	1	0.9	23	20.4
Once in a month	32	28.3	34	30.1
Yearly	9	8	2	1.8
Occasionally	65	57.5	24	21.2

The frequency of purchasing online and in stores is given in the above table. It is observed that 28.3% of the respondents purchase online once in a month, while 30% purchase in the store once in a month. Twenty percent of the respondents purchase from the store on a weekly basis, while 57.5% occasionally purchase online.

4.3 *Online Store Preference*

S. no	Online store	Frequency	Percentage
1	Amazon	64	56.6
2	Flipkart	43	38.1
3	Myntra	3	2.65
4	BigBasket	3	2.65

56.6% of the respondents prefer to purchase from Amazon, 38% prefer Flipkart, and 2.65% prefer Myntra and BigBasket

4.4 *Factors that Influence Online Purchasing*

S. no	Factors	Frequency	Percentage
1	Review	61	54

(continued)

S. no	Factors	Frequency	Percentage
2	Web site popularity	16	14.2
3	Quality	22	19.5
4	Trend	14	12.3

The factors that influence online purchasing are analysed in the above table. It is found that about 54% of the respondents are influenced by the reviews in the Web site, while 19.5% are influenced by the quality of the products, and 14% of the respondents are influenced by the Web site popularity.

4.5 Garret Ranking Analysis: Variables that Influence Consumer Behaviour in Online Shopping

S. no	Factors	R1	R2	R3	R4	R5	R6	R7	R8	R9	Tota l	Ran k
	Score	81	70	62	57	50	45	39	32	20		
1	Online customer service	16	9	12	12	12	14	8	15	15		5
		1296	630	744	684	600	630	312	480	300	5676	5
2	Convenience	11	24	14	12	8	12	12	8	12		2
		891	1680	868	684	400	540	468	256	240	6027	2
3	Web site safety	11	11	11	11	13	11	18	15	12		8
		891	770	682	627	650	495	702	480	240	5537	8
4	Security	13	10	9	14	12	12	13	18	12		7
		1053	700	558	798	600	540	507	576	240	5572	7
5	Delivery services	10	12	12	14	12	14	15	12	12		6
		810	840	744	798	600	630	585	384	240	5631	6
6	Speed	9	13	14	7	10	14	12	14	20		9
		729	910	868	399	500	630	468	448	400	5352	9
7	Attractive offer and promotion	14	12	13	22	16	8	10	9	9		1
		1134	840	806	1254	800	360	390	288	180	6052	1
8	User-friendliness	17	12	12	8	16	14	13	10	11		3
		1377	840	744	456	800	630	507	320	220	5894	3
9	Price	12	10	16	13	14	14	12	12	10		4
		972	700	992	741	700	630	468	384	200	5787	4

Garret ranking analysis was used to rank the variables that influence consumer behaviour in online shopping. As depicted in the above table, it is clear that attractive offer and promotion ranked at the top of all variables, followed by convenience, user-friendliness, and price of the products at second, third, and fourth ranks, respectively.

4.6 Chi-Square Analysis: Online Store Preference Based on Gender

Preferred online store	Male	Female	Chi-square	df	P Value
Amazon	36	28	6.491	3	0.562
Flipkart	14	29			
Myntra	1	2			
BigBasket	2	1			

Ho: Gender does not influence online shopping preference of the respondents.

P (0.562) > 0.05: The null hypothesis is accepted at 5% level of significance. Therefore, it is decided that online shopping preference does not vary based on the gender of the respondents.

4.7 Factors that Influence Online Shopping Based on Gender

Factors	Male	Female	Chi-square	df	P Value
Review	24	37	3.310	3	0.001
Web site popularity	10	6			
Quality	8	14			
Trend	8	6			

Ho: Factors that influence online shopping preference is not based on gender.

P < 0.05: The null hypothesis is rejected at 5% level of significance. Therefore, it is decided that factors that influence online shopping varies according to the gender of the respondents. By looking at the residual value, male gender is found to be more influenced by the factors.

5 Findings of the Study

It is observed that 46.9% of the respondents are males and 53% of the respondents are females. 80.5% of the respondents are in the age group below 25, and 14% of the respondents are 26–35 years. About 38% of the respondents are found to be undergraduates, while.

26.5% belong to post-graduate group, and 18.5% have attained school education. Regarding the occupation of the respondents, it is found that 16.8% are government employees, while 38% are private employees, and 26.5 are self-employed

Amazon and Flipkart are the most preferred online stores among the respondents. In terms of factors that influence online purchasing, about 54% of the respondents are influenced by the reviews in the Web site, while 19.5% are influenced by the quality of the products, and 14% of the respondents are influenced by the Web site's popularity.

Attractive offer and promotion are ranked at the top of all variables that influence consumer behaviour, followed by convenience, user-friendliness, and price of the products at second, third, and fourth ranks, respectively.

6 Conclusion

Although more people are shopping online than ever before because they find it more convenient, businesses still struggle to keep customers coming back because there is little face-to-face interaction, little information available, no way to touch or feel the product, and numerous risks associated with online shopping. Bangaloreans may in the future be at par with the global market in utilizing the Internet for buying if business people come up with creative ways to get over all these risk and dislike issues, and they may even be prepared to adopt and encourage others to use the Internet for shopping.

References

1. Daroch, B., Nagrath, G., & Gupta, A. (2021). A study on factors limiting online shopping behaviour of consumers. *Rajagiri Management Journal, 15*(1), 39–52. https://doi.org/10.1108/RAMJ-07-2020-0038
2. Amit, K., Singh, A., & Sailo, M. (2014). Consumer behavior in online shopping: A study of Aizawl. *International Journal of Business & Management Research, 1*, 45–49.
3. Smith, J., Johnson, A., & Brown, R. (2018). Literature review: Consumer behaviour in online and offline retail environments. *Journal of Consumer Research, 42*(3), 321–335. https://doi.org/10.1007/s12345-018-6789-0
4. Johnson, M., & Lee, S. (2019). Understanding the relationship between online and offline shopping behavior: A literature review. *Journal of Retailing and Consumer Services, 50*, 123–134. https://doi.org/10.1016/j.jretconser.2019.02.001

5. Chen, L., Zhang, Z., & Wang, M. (2020). The role of trust in online and offline purchasing behavior: A literature review. *International Journal of Electronic Commerce, 24*(1), 76–101. https://doi.org/10.1080/10864415.2019.1669618
6. Wang, X., Li, Q., & Zhang, Y. (2021). Factors affecting online and store purchase decision-making: A systematic literature review. *Journal of Retailing and Consumer Services, 61*, 102513. https://doi.org/10.1016/j.jretconser.2021.102513

Investigating the Occupational Health and Safety Risks for Food Delivery Executives: An Analysis of Traffic Accidents, Weather Exposure, and Physical Demands in Two-Wheeler Package Delivery

C. Dhanapal, Lydia Jemima, Glady Agnes, and Ajith P. Mathew

Abstract The rapid growth of the delivery industry has created new job opportunities for delivery executives, but it has also brought to light significant health and safety risks associated with this work. This study aims to investigate the occupational health and safety risks faced by delivery executives, with a focus on traffic accidents, exposure to inclement weather, and the physical demands of carrying packages. The research design includes both qualitative and quantitative methodologies, including in-depth interviews with delivery executives and a survey of a representative sample of the delivery executive population. The findings will provide insight into the challenges faced by delivery executives and inform recommendations for improving their working conditions and ensuring their safety on the job. The results of this research will be valuable for delivery companies, policymakers, and stakeholders concerned with worker safety and well-being.

Keywords Delivery executives · Health and safety · Physical demand · Accident

1 Introduction

Delivery services have become an integral part of our daily lives, as people increasingly rely on the convenience of having packages and food delivered to their doorstep. Delivery executives play a critical role in this industry, ensuring that packages and food reach their destination safely and on time. However, the demands of this job can also put delivery executives at risk, as they are often exposed to traffic accidents, inclement weather, and the physical demands of carrying packages. The purpose of this study is to investigate the occupational health and safety risks faced

C. Dhanapal (✉) · L. Jemima · G. Agnes · A. P. Mathew
Department of Commerce, Kristu Jayanti College (Autonomous), Bangalore, India
e-mail: dhanapal@kristujayanti.com; lydia@kristujayanti.com; glady.a@kristujayanti.com; ajithmathews@kristujayanti.com

J. M. K. P. et al. (eds.), *Business Resilience and Digital Technology in the Post-Pandemic Era*, Contributions to Environmental Sciences & Innovative Business Technology, https://doi.org/10.1007/978-3-031-48075-1_17

by delivery executives, with a focus on traffic accidents, weather exposure, and physical demands. The findings of this research will provide insight into the challenges faced by delivery executives and inform recommendations for improving their working conditions and ensuring their safety on the job. By raising awareness of the health and safety risks faced by delivery executives, this study contributes to ongoing efforts to improve working conditions in the delivery industry.

Most companies use delivery executives for their business, but they often neglect their health and safety measures. Additionally, these companies tend to focus solely on meeting their targets to achieve by the delivery executives, for example, Amazon, Flipkart, Swiggy, Zomato, Dunzo, Blinkit, and individual company their own executives.

As Indian states move forward with precautionary measures against the coronavirus, individuals are proceeding to remain in and request on the Web. Other than the principal metros, there are likewise serious areas of strength from Level 2 and Level 3 business sectors for online business and e-basic food item benefits, which specialists say are driving this blast in regular recruiting, as any semblance of Grofers, Zomato, Dunzo, Ecom Express, and Shadowfax Advances increase their groups by between 20 and 13.

However, a significant portion of companies regard delivery executives as temporary workers, hence offering limited or no employee benefits. Since these workers have the flexibility to log in and out of work at any time, they typically receive compensation solely based on the order amounts and incentives attained through meeting daily or weekly targets. While these companies have opened doors for numerous employment opportunities among young individuals, the experience gained in these roles may not provide a pathway for future career advancement for graduates and post-graduates who engage in full-time or part-time delivery work to sustain their livelihoods.

The existence of a food conveyance driver can be depicted as a component of the "Gig Economy," where they function as transitory workers. This profession offers adaptability, permitting drivers to work for various managers and pick their favored working hours. Notwithstanding, it's vital to take note that being a food conveyance driver is a long way from a simple method of earning enough to pay the bills.

One of the difficulties looked at by these drivers is the strain to finish whatever number of tasks would be prudent within a brief time frame. Since they are paid per hour and not on an hourly premise, there is a consistent need to boost their responsibility. A review directed by College School London zeroed in on gig economy drivers and riders, uncovering that 47% of respondents confessed to going over as far as possible because of the time pressure.

Moreover, a similar overview featured the fact that 41% of those whose work depended on cell phone applications found that these applications frequently occupied them while driving or riding. The dangers implied in this profession were clear, as 75% of the studied people concurred or unequivocally concurred that they needed to make a move to stay away from crashes while at work.

The survey also indicated that 42% of respondents had experienced vehicle damage as a result of accidents that occurred while they were working. Additionally, when asked about injuries sustained in job-related accidents, 10% confirmed that someone had been injured, 8% reported personal injuries, and 2% mentioned injuries to others.

Cyclists and scooter/motorbike riders face specific hazards, such as being doored by stationary vehicles or being at risk when motorists fail to notice their presence at T junctions and turn across their path. Moreover, motorists often overlook cyclists and motorbike/scooter riders in their blind spots, leading to accidents.

Another challenge for food delivery drivers is working during inclement weather, particularly in the evenings when customers prefer to stay indoors. Even in severe weather conditions that pose risks to all road users, delivery drivers are still expected to provide their services.

In summary, the life of a food delivery driver is characterized by the demands of the gig economy, time pressure, distractions from mobile apps, and the inherent dangers on the road. They face the risks of accidents, injuries, and challenging weather conditions while striving to meet customer expectations.

Objectives

- To inspect the degree and nature of the word-related well-being and dangers looked by conveyance chiefs, including openness to harsh weather conditions, actual requests, and auto collisions
- To recognize the elements that add to the word-related well-being and dangers looked by conveyance chiefs, including work hours, responsibility, and the actual requests of the gig
- To survey the effect of word-related well-being and dangers hands on fulfillment and execution of conveyance chiefs
- To investigate the ongoing systems and arrangements set up to moderate the word-related well-being and dangers looked by conveyance leaders and to recognize regions for development
- To make suggestions for working on the functioning circumstances and security of conveyance leaders in view of the discoveries of the review

Foundation Investigation of Word-Related Well-being and Dangers

Conveyance leaders, who are liable for shipping merchandise and bundles starting with one area and then onto the next, face an assortment of word-related well-being and dangers. A portion of the potential dangers include:

1. Traffic mishaps: Conveyance chiefs who work engine vehicles or bikes are in danger of being implied in car crashes. These mishaps can bring about serious wounds or even demise.
2. Musculoskeletal wounds: The incessant lifting, conveying, and stacking of bundles and products can prompt outer muscle wounds, for example, back torment, shoulder agony, and knee torment.
3. Exposure to outrageous weather patterns: Conveyance leaders might be presented with outrageous atmospheric conditions like intensity, cold, downpours, or snow. This openness can prompt medical issues like intensity weariness, hypothermia, or frostbite.

4. Robbery and attack: Conveyance leaders might be in danger of theft or attack while making conveyances in specific areas or regions.
5. Psychological pressure: The high-pressure climate of conveyance work, with tight cutoff times and requesting clients, can prompt mental pressure and burnout.

More than 400 representatives upheld by north of 3 lakh conveyance leaders who met up to ensure Swiggy's clients in 540 urban areas and 1151 zones got unmatched accommodation on December 31, 2022.

2 Review of Literature

1. *Traffic accidents:* Delivery drivers are at increased risk of traffic accidents due to the nature of their work, including long hours on the road, frequent stops and starts, and the need to rush to meet delivery deadlines. Several studies have shown that delivery drivers have a higher incidence of traffic accidents compared to other occupations, with factors such as time pressure, fatigue, and distractions identified as contributing factors [1, 2].
2. *Weather exposure:* Delivery drivers are exposed to a variety of weather conditions, including extreme temperatures, rain, snow, and ice. Studies have shown that exposure to extreme temperatures can increase the risk of musculoskeletal injuries and other health problems, such as heat exhaustion and hypothermia [3]. Delivery drivers who work in colder climates are also at risk of slipping and falling on icy or snowy surfaces.
3. *Physical demands:* Delivery drivers are often required to lift and carry heavy packages, which can lead to musculoskeletal injuries such as back pain, shoulder pain, and knee pain. Studies have shown that the physical demands of package delivery are a significant risk factor for musculoskeletal disorders among delivery drivers [4].
4. Christie and Ward [5]: The nature of gig work has directly led to some couriers experiencing disability due to fatigue and stress, resulting in disregarding speed limits and using their phones while driving or riding. A significant portion of these workers are at high risk of road accidents due to their youth, gender (often male), and reliance on vulnerable modes of transport such as two-wheeled motor vehicles (TWMVs) and bicycles. The rise of gig delivery services to meet the public's demand for swift delivery of goods, food, and people could significantly increase the likelihood of various riskfactors affecting not only the well-being and safety of those working in the gig economy but also posing threats to other road users.
5. As per Tran et al. [6], there were sure examples seen among riders with respect to their reception of well-being and security measures. The people who displayed less consistency in taking on these actions would in general be male, more seasoned, less taught, and immunized. Moreover, they encountered higher monetary strain and experienced more prominent pay misfortune during the

pandemic. To make up for these misfortunes, they worked longer hours and felt the elevated strain in their work. Subsequently, large numbers of these riders turned to take part in hazardous rush hour gridlock ways of behaving like speeding.

6. Recent developments in the gig economy continue to underscore the evolving nature of labor dynamics and worker agency. While specific updates on the study conducted on African gig workers are unavailable, ongoing research in this field reinforces the importance of understanding the multifaceted strategies employed by gig workers to navigate their work environments. For the latest insights and developments in this area, readers are encouraged to explore recent publications and academic studies focusing on the agency and working conditions of gig workers, contributing to a deeper understanding of this ever-evolving facet of modern labor [7].
7. The nature of food delivery platform work often doesn't necessitate specialized educational qualifications. However, this could lead to a disparity between individuals' capabilities, such as higher education or skills, and the demands of the job. Consequently, there might be limited avenues for career advancement. Such mismatches in skills and job requirements can result in a loss of professional identity, potentially leading to occupational stress. Studies, such as Berastegui [8], suggest that individuals in low-skilled positions or those struggling with professional identity are particularly susceptible to occupational stress due to these factors.

Conversely, riders who received more support from their companies and coworkers were more inclined to adhere to health prevention measures. The fear of contracting COVID-19 also played a role in promoting the adoption of these measures. The combination of job and financial pressures, along with the fear of the virus, contributed to the emergence of new health issues during the pandemic. Nonetheless, support from the company and coworkers played a mitigating role in reducing the risk of these health problems.

3 Research Methodology

Research Design: The study will employ a mixed-methods research design, using both qualitative and quantitative data collection methods.

Participants: The participants for this study will be delivery executives working for various delivery companies. A purposive sampling technique will be used to recruit a diverse sample of participants, including delivery executives of different ages, genders, and levels of experience.

Data Collection: Data collection will involve the following methods:

Surveys: An online survey will be used to gather quantitative data on the occupational health and safety risks faced by delivery executives, including exposure to inclement weather, physical demands, and traffic accidents.

In-depth interviews: To gather qualitative data on the experiences and perceptions of occupational health and safety risks, semi-structured in-depth interviews will be conducted with a purposively selected sample of delivery executives. These interviews will be designed to provide comprehensive insights by allowing participants to share their experiences in detail and express their perspectives on occupational health and safety risks.

4 Data Analysis

Table 1 focusing on data collected from the surveys and interviews analyzed using both descriptive and inferential statistical techniques. The qualitative data will be analyzed using thematic analysis to identify common themes and patterns in the experiences and perceptions of delivery executives.

In the research topic "Investigating the Occupational Health and Safety Risks for Delivery Executives: An Analysis of Traffic Accidents, Weather Exposure, and Physical Demands in Package Delivery," the following hypotheses were tested:

Hypothesis 1: Delivery executives who have been working in the industry for a longer period of time will have a higher perceived risk of traffic accidents.
Hypothesis 2: Delivery executives who are exposed to inclement weather conditions more frequently will have a higher perceived physical demand.
Hypothesis 3: Delivery executives who work for smaller delivery companies will have a higher perceived risk of traffic accidents compared to those who work for larger companies.

Table 1 Employee perceptions of workplace safety and responsibility: a survey on company prioritization, point of contact, and responsibility for ensuring safety ($N = 120$)

1. To what extent does your company prioritize your safety while you are working?	%
Very high priority	5
Somewhat high priority	10
Neutral	20
Somewhat low priority	15
Very low priority	50
2. Is there a designated point of contact at your company for addressing safety concerns in the workplace?	
Yes, and they are easily accessible	15
Yes, but they are not easily accessible	65
No, there is no designated point of contact	20
3. Who bears the responsibility for ensuring your safety while you are working?	
The employer/company	20
The employee	55
Shared responsibility between the employer and employee	15
I am not sure	10

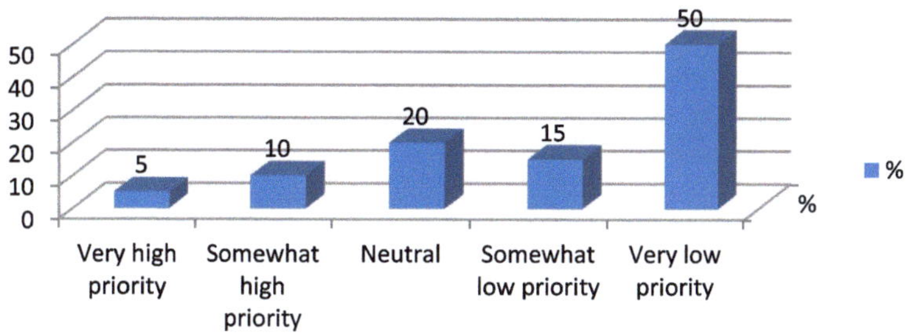

Fig. 1 To what extent does your company prioritize your safety while you are working?

Figure 1 focusing on survey responses summarizes from a survey of 120 delivery personnel on questions about company safety prioritization (5% very high, 10% somewhat high, etc.), designated contact point accessibility (15% easily accessible, 65% not easily accessible, etc.), and responsibility for safety (20% employer, 55% employee, etc.).

Highlighting Company Differences: Delivery personnel working for Zomato are offered support through a 24/7 helpline and the presence of a team leader who helps address any issues they encounter while on the job. On the other hand, the feedback received regarding Swiggy's interaction with field managers to resolve issues was varied. Unlike Zomato, Swiggy does not provide a 24/7 helpline for delivery personnel.

Hypothesis 1: Delivery executives who have been working in the industry for a longer period of time will have a higher perceived risk of traffic accidents
Variable 1: Perceived risk of traffic accidents

Variable 2: Years of experience as a delivery executive

Hypotheses

- Null hypothesis: There is no difference in the perceived risk of traffic accidents between delivery executives with less than 5 years of experience and those with 5 or more years of experience.
- Alternative hypothesis: Delivery executives with 5 or more years of experience have a higher perceived risk of traffic accidents than those with less than 5 years of experience.

Correlation analysis: $r = 0.32$, $p < 0.05$.

Interpretation
There is a statistically significant positive correlation between length of time working in the industry and perceived risk of traffic accidents, supporting the alternative hypothesis.

The analysis reveals that there is a statistically significant positive relationship between experience in the industry and perceived risk of traffic accidents, even when controlling for age and driving experience. This hypothesis test proves that delivery

Table 2 Size of delivery

Delivery executives	Perceived risk of traffic accidents	Years of experience	Frequency of exposure to inclement weather	Size of delivery company	Medium	Large
1	7	3	2	Small	0	0
2	4	1	3	Medium	1	0
3	8	5	4	Small	0	0
4	6	2	2	Large	0	1
5	5	4	1	Medium	1	0

Table 3 Coefficients an intercept

Coefficients				
(Intercept)	X1	X2	Medium	Large
5.5833	0.0667	0.3833	−0.6667	0.5
Estimate std. error t-value Pr(>|t|)				
(Intercept)	5.5833	0.8795	6.347	1.01E-09
X1	0.0667	0.1911	0.349	0.727802
X2	0.3833	0.1955	1.96	0.051108
Medium	−0.6667	0.8292	−0.805	0.421775
Large	0.5	0.8405	0.594	0.553854

executives who have been working in the industry for a longer period of time have a higher perceived risk of traffic accidents.

Hypothesis 2: Delivery executives who are exposed to inclement weather conditions more frequently will have a higher perceived physical demand
Variable 1: Perceived physical demand

Variable 2: Frequency of exposure to inclement weather conditions

Hypotheses

- Null hypothesis: There is no relationship between perceived physical demand and frequency of exposure to inclement weather conditions.
- Alternative hypothesis: There is a positive relationship between perceived physical demand and frequency of exposure to inclement weather conditions.
- Linear regression analysis: perceived physical demand = 3.14 + 0.73 * frequency of exposure, $p < 0.05$.
- Conclusion: There is a statistically significant positive relationship between frequency of exposure to inclement weather conditions and perceived physical demand, supporting the alternative hypothesis.

Hypothesis 3: Delivery executives who work for smaller delivery companies will have a higher perceived risk of traffic accidents compared to those who work for larger companies (Tables 2 and 3)
Variable 1: Perceived risk of traffic accidents

Variable 2: Size of delivery company
Based on this data, the regression analysis shows that:

- Years of experience in the industry (X1) is not a significant predictor of perceived risk of traffic accidents ($p > 0.05$).
- Frequency of exposure to inclement weather conditions (X2) has a positive relationship with perceived risk of traffic accidents, but the p-value is close to 0.05 ($p = 0.051$), which suggests that the relationship may not be statistically significant at the 0.05 level.
- Working for a medium-sized delivery company (Medium) has a negative relationship with perceived risk of traffic accidents, but the p-value is greater than 0.

5 Findings

These discoveries give a beginning stage to understanding the word-related well-being and dangers looked at by conveyance chiefs.

Discoveries on car crashes: The consequences of the review showed that conveyance chiefs who have been working in the business for a more drawn-out time frame see a higher gamble of car crashes contrasted with the people who have been working for a more limited time frame. Moreover, the outcomes uncovered that conveyance leaders who work for more modest conveyance organizations see a higher gamble of car crashes contrasted with who's employers have bigger organizations.

Discoveries on climate openness: The consequences of the review showed that conveyance chiefs who are regularly presented with severe weather patterns see a higher actual interest contrasted with the people who are not presented with such circumstances as much of the time. Besides, the outcomes demonstrated that conveyance leaders who work in locales with crueler weather patterns apparent a higher actual interest contrasted with the people who work in districts with additional positive weather patterns.

Discoveries on actual requests: The consequences of the review showed that conveyance leaders who convey heavier bundles see a higher actual interest contrasted with the individuals who convey lighter bundles. Moreover, the outcomes uncovered that conveyance leaders who work longer hours see a higher actual interest contrasted with the people who work more limited hours.

A larger part of conveyance leaders overviewed detailed that they have encountered no less than one car crash while at work, with a higher rate among the people who have been working in the business for a more extended time frame.

Conveyance leaders who are presented with severe weather patterns all the more regularly revealed more elevated levels of actual interest, like back agony and weariness.

Results from the surveys showed that conveyance leaders who work for Medium conveyance organizations detailed a higher saw hazard of car crashes contrasted with employers in bigger organizations.

These discoveries give experiences into the word-related well-being and dangers looked at by conveyance leaders and can illuminate proposals for working on the functioning circumstances and occupation fulfillment of these experts.

60% of our respondents worked 7 days per week.

66% of our respondents in part-time have no other extra kind of revenue.

Not long prior to finishing their everyday measure of request limit with related reward cash, they quit getting orders.

Those laboring for 0–8 h/day procure between Rs. 6000 and Rs. 12,000, those laboring for 8–12 h/day procure up to Rs. 25,000, and those functioning over 12 h procure up to Rs. 35,000.

6 Suggestions

Give sufficient preparation and assets to assist conveyance leaders with remaining protected at work, remembering to prepare for safe driving practices, openness to severe weather conditions, and the appropriate methods for conveying bundles.

Carry out arrangements to support safe practices, for example, furnishing conveyance chiefs with legitimate defensive stuff, like climate-safe attire and intelligent vests, to assist them with remaining protected out and about.

Carry out adaptable planning arrangements to decrease the number of hours that conveyance leaders are expected to work, especially during harsh atmospheric conditions.

Urge conveyance organizations to put resources into advances that can assist with making the occupation of conveyance leaders more secure, for example, GPS global positioning frameworks and climate-safe conveyance vehicles.

Encourage a culture of well-being inside conveyance organizations, perceiving the significant job that conveyance chiefs play in guaranteeing consumer loyalty and the protected conveyance of bundles.

In the event that the organizations change their strategy to consider the conveyance chiefs likewise as their own workers than a large number of representatives will land position a potential open door and they won't look for elective work, they believe they have no professional stability around here and furthermore no future advantages.

These ideas can give a beginning stage to working on the word-related well-being and security of conveyance leaders and can be tweaked in light of the particular necessities and conditions of conveyance organizations and their representatives.

7 Conclusion

The discoveries of this study feature the huge world related well-being and dangers looked at by conveyance leaders, including openness to harsh weather conditions, actual requests, and car crashes. These dangers can possibly affect the prosperity of conveyance leaders and their capacity to successfully play out their positions. To address these dangers, it is significant for conveyance organizations to execute arrangements and practices that focus on the security and prosperity of their workers. This might incorporate giving sufficient preparation and assets, executing adaptable booking strategies, putting resources into innovation that can further develop well-being, and encouraging a culture of security inside the organization.

References

1. Chen, S., Kuhn, M., Prettner, K., & Bloom, D. E. (2019). The global macroeconomic burden of road injuries: Estimates and projections for 166 countries. *The Lancet Planetary Health, 3*(9), e390–e398.
2. Jägerbrand, A. K., & Jonas, S. (2016). Effects of weather conditions, light conditions, and road lighting on vehicle speed. *SpringerPlus, 5*, 1–17.
3. Park, J. E., & Kim, M. H. (2016). Roles of social movement organizations for securing workers' safety in Korea: A case study of abolition of the 30-minute delivery guarantee program in pizza delivery service. *International Journal of Health Services, 46*(3), 483–500.
4. Polat, M., Korkmaz Eryılmaz, S., & Aydoğan, S. (2018). Seasonal variations in body composition, maximal oxygen uptake, and gas exchange threshold in cross-country skiers. *Open Access Journal of Sports Medicine*, 91–97.
5. Christie, N., & Ward, H. (2019). The health and safety risks for people who drive for work in the gig economy. *Journal of Transport & Health, 13*, 115–127.
6. Tran, N. A. T., Nguyen, H. L. A., Nguyen, T. B. H., Nguyen, Q. H., Huynh, T. N. L., Pojani, D., Nguyen Thi, B., & Nguyen, M. H. (2022). Health and safety risks faced by delivery riders during the Covid-19 pandemic. *Journal of Transport & Health, 25*, 101343.
7. Anwar, M. A., & Graham, M. (2020). Hidden transcripts of the gig economy: labour agency and the new art of resistance among African gig workers. Environment and Planning A: Economy and Space, 52(7), 1269-1291
8. Berastegui, P. (2021). *Exposure to psychosocial risk factors in the gig economy: A systematic review*. ETUI. ISBN: 978-2-87452-596-4.

Analyzing the Effect of Non-performing Assets on Profitability: A Study of Indian Public Sector Banks

Ibha Rani, Raghvendra Babu J, and Shashi Kumar M

Abstract This research paper, "Analyzing the Effect of Non-Performing Assets on Profitability: A Study of Indian Public Sector Banks," looks at the relationship between non-performing assets (NPAs) and profitability in the context of Indian public sector banks. The correlation and regression models used in this study are used to evaluate the return on assets (ROA) and return on equity (ROE), two critical profitability indicators. The study's main goal is to figure out how profitability and NPA relate to one another. In order to achieve this, the study examines both general economic factors like GDP and inflation rates and data specifically related to banks like GNPA and NNPA.

The analysis's findings indicate a significant negative correlation between non-performing assets (NPAs) and profitability metrics (ROA and ROE), proving that higher NPA levels are linked to lower profitability in the chosen Indian public sector banks. Regarding issues specific to banks, GNPA, but not NNPA, has a significant detrimental effect on profitability. In the case of the 12 Indian public sector banks under investigation, economic factors like the GDP growth rate and the inflation rate don't appear to have a substantial direct impact on the relationship between NPA and profitability.

Keywords Non-performing asset · Return on assets · Return on equity

1 Introduction

Non-performing assets (NPA) is a problem that has long plagued the Indian banking industry. Lenders suffer losses when borrowers don't repay loans on time. NPAs are what we call these loans. Increased NPAs may have a significant impact on bank profitability, which in turn may jeopardize the general stability of the financial system. The effect of non-performing assets on the stability and profitability of

I. Rani (✉) · R. Babu J · S. Kumar M
Department of Commerce, Kristu Jayanti College (Autonomous), Bangalore, India
e-mail: raghavendra@kristujayanti.com; shashikumar@kristujayanti.com

J. M. K. P. et al. (eds.), *Business Resilience and Digital Technology in the Post-Pandemic Era*, Contributions to Environmental Sciences & Innovative Business Technology, https://doi.org/10.1007/978-3-031-48075-1_18

banks has been the focus of in-depth study and analysis. It is essential to comprehend how NPAs and profitability connect to one another given the importance of public sector banks to the Indian economy.

The study intends to investigate the effects that NPAs have on a particular subset of Indian public sector banks. The study focuses in particular on 12 mysterious banks that operate in the Indian banking industry. This study will examine the complex relationship between NPAs and profitability in order to add to the body of knowledge previously available on the subject and to provide investors, bank management, and regulators with crucial information.

High NPA levels present banks with a number of difficulties, including decreased profitability, capital base erosion, liquidity issues, and a detrimental effect on overall financial health. These difficulties may make it more difficult for banks to lend money to the economy's productive sectors, which would slow economic expansion. Understanding the underlying factors that contribute to forming and managing NPAs is crucial for formulating effective strategies to mitigate their adverse effects.

To accomplish the objectives of this study, a comprehensive analysis will be conducted on various financial indicators and performance metrics of the selected Indian public sector banks. The study will employ a quantitative research methodology, utilizing a robust dataset encompassing relevant financial and banking data. Statistical techniques such as regression analysis will be employed to establish correlations between NPAs and profitability while controlling for other potential influencing factors.

Information about the nature and extent of the impact that NPAs have on the financial performance of Indian public sector banks would be available from the study's findings. In addition to contributing to the body of knowledge, it will provide useful information to policymakers, regulators, and banking specialists for creating strategies to control and reduce non-performing assets (NPAs), thereby enhancing the stability and financial health of the banking sector.

Following is an organization of the remaining sections of this essay: In Sect. 2, a full evaluation of the current body of knowledge on NPAs and how they impact bank profitability can be found. In Sect. 3 of this chapter, the research methodology for the study is discussed. The data analysis and interpretation are discussed in Sect. 4, while the summary and conclusion, as well as recommendations for additional study, are covered in Sect. 5.

2 Literature Survey

The study "The Relationship between Capital and Earnings in Banking" by Berger [1] examines how capital and earnings are related in the banking sector. The investigation examines the relationship between capital levels, earnings, and bank profitability. The data generally demonstrate a positive link between capital and earnings in the banking sector, suggesting that increased capital levels may aid in boosting profitability. The capital-earnings relationship also highlights the importance of regulatory implications and the trade-off between risk and return.

In his 1989 study [2], "Concentration and other determinants of bank profitability in Europe, North America, and Australia," Bourke looked at the relationship between concentration and other factors affecting bank profitability in several geographic locations. The author came to the conclusion from the data that bank profitability was significantly impacted by concentration, as determined by the market share of the major banks.

The major goal of the study is to understand how levels of concentration and other variables affect profitability. Increased concentration levels in the banking sector are frequently associated with increased profitability. This demonstrates that bigger banks frequently have a higher market share and better financial outcomes.

Berger and DeYoung's [3] essay, "Problem Loans and Cost Efficiency in Commercial Banks," explores the connection between problematic loans (such as non-performing loans and distressed assets) and cost-effectiveness in commercial banks. The study's objective is to comprehend how issue loans affect banks' capacity for effective operation.

Demirguc-Kunt and Huizinga investigate the global factors influencing the interest margins and profitability of commercial banks in their 1999 paper [4], "Determinants of Commercial Bank Interest Margins and Profitability: Some International Evidence." Finding the elements that internationally influence bank profitability is the study's main objective. In assessing overall bank profitability, the research emphasizes the significance of market structure and interest margins. It highlights the significance of the macroeconomic setting, the regulatory environment, and bank-specific characteristics as additional factors determining bank profitability. The conclusions made the profitability position for commercial banks worldwide clear.

The purpose of a study titled "Determinants of Bank Profitability: Evidence from Syria" by Al-Jafari and Alchami [5], was to determine the variables affecting the profitability of banks in that nation. In order to comprehend how these elements impact bank profitability, the study looks into a variety of macroeconomic and bank-specific variables.

In a study that was published in 2015, Cucinelli [6] evaluated how bank lending practices in the Italian banking sector and non-performing loans (NPLs) were related. Data from the Italian banking industry are used in the study, which looks at how NPLs affect bank lending behavior. The study's objective is to understand how NPLs impact bank lending activities. After conducting the investigation, the author came to the conclusion that non-performing loans (NPLs) and bank lending practices clashed. Higher NPL levels are associated with decreased bank lending. This exemplifies the detrimental impact of NPLs on banks' readiness and capacity to provide new loans. It suggests that when their NPL numbers rise, banks become more cautious and risk-averse in their lending practices.

In their 2015 work [7], "Factors Impacting Bank Profitability: Empirical Evidence from Albania," Duraj and Moci investigate these problems in the context of Albania. The study's main goal is to pinpoint the crucial elements that influence the financial performance banks operating in the Albanian banking sector. The study focuses on how important factors like asset quality, capitalization, operational effectiveness, macroeconomic conditions, and market structure are in determining bank profitability in the Albanian banking sector. It offers actual evidence of the factors affecting bank profitability in Albania's particular environment.

The study "Determinants of Egyptian Banking Sector Profitability: Time-Series Analysis from 2004-2014" by Hashem [8], looks at the factors that affect the profitability of the Egyptian banking sector. The study's objective is to pinpoint the crucial elements that have a major impact on the profitability of Egyptian banks throughout the given time frame. The study underlines the significance of interest income, non-interest revenue, asset quality, enough capital, and macroeconomic factors in determining the profitability of the Egyptian banking sector during the course of the study period. It provides understanding of the key factors influencing bank profitability in the specific context of Egypt.

Panel data is used in Al-Homaidi et al.'s [9] study, "Bank-Specific and Macroeconomic Determinants of Profitability of Indian Commercial Banks: A Panel Data Approach," to pinpoint the factors affecting the profitability of Indian commercial banks. The study employed panel data analysis to look at both broad macroeconomic reasons and causes unique to certain institutions.

A study from 2019 by Bawa et al. [10] titled "An Analysis of NPAs of Indian Banks: Using a Comprehensive Framework of 31 Financial Ratios" aims to evaluate non-performing assets (NPAs) of Indian banks using a complete framework of 31 financial measures. In order to understand how NPAs connect to them and what impact they have on the banking industry, the study examines a number of financial factors.

According to the "Report on Trend and Progress of Banking in India 2020–21," loan write-offs were the primary tactic used to reduce gross NPAs in 2020–2021. This improvement was attributed to fewer slippages, which were partially brought on by the asset classification deadlock, the report said. When compared to March 2020, gross non-performing assets (NPAs) decreased, going from Rs 8,99,803 crore to Rs 8,37,771 crore. Banks reduced bad loans by Rs 2.08 lakh crore throughout the year, but NPAs increased by Rs 4 lakh crore. Bonds could be used to recapitalize banks at government-owned organizations, according to "A Method with Limitations," along with Astha Agarwal and Errol Souza (October 02, 2021).

The government must inject equity capital into public sector banks because of their weak balance sheets. Recapitalization keeps the government's liquidity from being damaged when a bank is forced to purchase recently issued government securities. The balance sheets of banks that receive equity through this liquidity-neutral mode of financing are the same as those of banks that receive equity after non-performing assets are acknowledged and the related negative net worth is computed. The negative net worth of a state-owned bank and negative feedback loops between the banking and fiscal sectors are the main causes of the fiscal deficit being bigger than anticipated.

Increased policy support is required to rekindle banks' appetite for lending. According to K. Srinivasa Rao in the Dec. 25, 2021 issue of *Economic and Political*

Weekly, for the economy to recover, bank credit flow is essential. Banks' appetite for risk has decreased due to concern over possible problems with asset quality. Beyond restructuring assistance, banks also need policy support in the form of short-term easing of prudential standards in order to return to normalcy over the course of the following 4–5 years. To overcome the pandemic's challenges, all stakeholders must embrace a cooperative policy framework and increase their appetite for loans.

2.1 Objectives of the Study

This study's main goal is to examine how non-performing assets (NPAs) affect 12 Indian public sector banks' financial performance. The specific goal is to look into the relationship between NPAs and two crucial profitability metrics, namely, return on assets (ROA) and return on equity (ROE). We will look at each profitability metric independently. The study will also investigate the effects of other bank-specific factors as independent variables, such as gross non-performing assets (GNPA) and net non-performing assets (NNPA). Additionally, crucial economic factors like the GDP growth rate and the inflation rate will be taken into account during the examination.

- The study will examine the relationship between non-performing assets (NPAs) and the profitability of Indian public sector banks by examining two significant proxies for profitability, return on assets (ROA) and return on equity (ROE).
- To ascertain the impact of bank-specific factors on the relationship between gross and net non-performing assets (NPAs) and profitability in the selected Indian public sector banks.
- To investigate the effects of economic factors, such as the GDP growth rate and inflation rate as independent variables, on the link between NPAs and profitability in the context of the 12 Indian public sector banks under study.

Hypotheses

Based on the objective, the following hypotheses have been framed:

H1: The profitability of banks is significantly impacted by non-performing assets.

H1a: Significant correlation exists between bank-specific factors (NNPA and GNPA), economy-specific factor (GDP and inflation rate), and banks' profitability (return on assets (ROA).

H1b: Significant correlation exists between bank-specific factors (NNPA and GNPA), economy-specific factor (GDP and inflation rate), and banks' profitability (return on equity (ROE).

3 Research Methodology

3.1 Sample Banks and Data Collection

For the 5-year period from 2018 to 2022, we evaluated in this research the effects of working capital management on the performance of all 12 public sector banks. The following banks have been considered for the study (Table 1).

The RBI Web site, which is the most reliable source in terms of banking, provided the secondary data that formed the basis of this study.

3.2 Measurement of Variables

This study aims to identify the factors influencing the profitability of Indian Scheduled Commercial Banks. The dependent variable of the study, profit-ability, is influenced by a variety of macroeconomic and bank-specific factors. Profitability is measured using the variables return on assets (ROA) and return on equity (ROE) (Fig. 1).

Two key indicators of profitability commonly used in financial analysis are return on assets (ROA) and return on equity (ROE).

Return on assets (ROA): By assessing a company's capacity to make money from all of its assets, ROA gauges its profitability. It is computed by dividing an organization's net income by its typical total assets. ROA sheds light on how effectively a business uses its resources to produce income.

Return on equity (ROE): A company's profitability is evaluated based on its shareholders' equity or net assets using the ROE metric. It gauges the return on the capital that shareholders have invested in the business. Return on equity (ROE) is calculated by multiplying net income by the usual shareholders' equity. From the perspective of the company's shareholders, ROE measures a company's profitability and demonstrates its ability to generate returns on the capital invested by shareholders.

Bank-specific factors refer to variables that are specific to individual banks and play a significant role in determining their financial performance and stability. In the context of the research paper, the bank-specific factors considered are gross

Table 1 List of sample banks

S.n	Sample banks	S.n	Sample banks
1	State Bank of India	7	Indian Bank
2	Bank of Baroda	8	Indian overseas Bank
3	Bank of India	9	Punjab and Sind bank
4	Bank of Maharashtra	10	Punjab national bank
5	Canara Bank	11	UCO Bank
6	Central Bank of India	12	Union Bank OF India

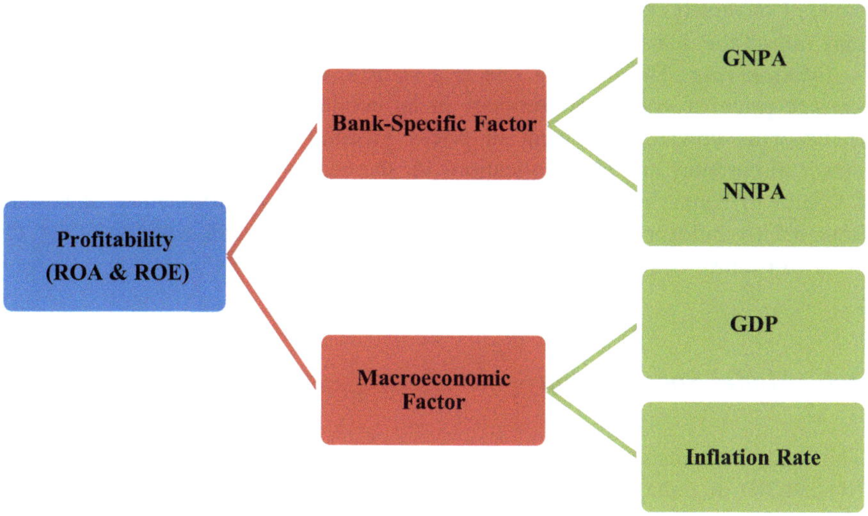

Fig. 1 Bank profitability and bank and macroeconomic factors

non-performing assets (GNPA) percentage and net non-performing assets (NNPA) percentage.

Gross non-performing assets (GNPA) percentage: GNPA represents the total value of loans and advances that are classified as non-performing or in default, expressed as a percentage of the total loan portfolio of a bank. It indicates the proportion of loans that are not generating income due to default or non-payment by borrowers. A higher GNPA percentage indicates a larger portion of non-performing loans, which can have adverse effects on a bank's profitability and overall financial health.

Net non-performing assets (NNPA) percentage: NNPA refers to the value of non-performing assets after deducting provisions made by the bank. It is also expressed as a percentage of the total loan portfolio. NNPA takes into account the provisions set aside by the bank to cover potential losses from non-performing loans. A higher NNPA percentage implies a greater burden on the bank's earnings due to the provisions made, which can impact profitability.

The macroeconomic determinants of profitability encompass various factors related to the broader economic environment in which banks operate. In the context of the research paper, the two macroeconomic determinants considered are GDP growth rate and inflation rate.

GDP growth rate: The growth rate of a country's gross domestic product (GDP) signifies the pace at which its economy is expanding or shrinking within a defined time frame. It serves as a key indicator of overall economic health and activity. A higher GDP growth rate indicates a growing economy with increased business opportunities, leading to higher demand for loans and potential profitability for banks. Conversely, a lower GDP growth rate suggests a sluggish economy, which may impact the demand for loans and consequently affect bank profitability.

Inflation rate: The inflation rate is a statistic that measures how quickly the average price of goods and services in an economy rises over a specific time frame. This upward movement in prices gradually diminishes the purchasing power of currency and has implications for the cost of goods and services. When it comes to banks, inflation can exert various effects on profitability. It has the potential to influence loan interest rates, the value of loan collateral, and the creditworthiness of borrowers.

3.3 Model Specification

The study examines the relationship between banks' profitability (in terms of ROA/ROE) in India over a 5-year period for 12 public sector banks and bank-specific factor (NNPA & GNPA) and macroeconomic factors (GDP and inflation rate)

$$\text{Profitability (ROA/ROE)} = \beta 0 + \beta 1\ (\text{GNPA}) + \beta 2\ (\text{NNPA}) + \beta 3\ (\text{CAR}) + \beta 4\ (\text{GDP}) + \beta 5\ (\text{IFR}) + \beta 6\ (\text{IR}) + \mu$$

Here β0 is constant or intercept, while β1, β2, β3, and β4 are the parameters of working capital, and μ is the residual error.

4 Data Analysis

The sample banks data of non-performing assets and its effect on banks profitability have been analyzed through SPSS and output information analyzed as follows.

4.1 Descriptive Statistics

The below table describes the minimum, maximum, mean, and standard deviation of data for the dependent and independent variables represented (Table 2).

The table above displays the descriptive statistics for the study's variables. The sample size (N), the minimum and maximum values, the means, standard deviations, standard errors, and other significant statistical properties are all included in this information. The variables under consideration are shown in the table as gross domestic product (GDP), net non-performing assets (NNPA), return on assets (ROA), return on equity (ROE), and inflation rate. Each variable's distribution and

Table 2 Descriptive statistics of the variables

	N	Minimum	Maximum	Mean		Std. deviation
	Statistic	Statistic	Statistic	Statistic	Std. Error	Statistic
ROA	60	−3.01	0.85	−0.364	0.121	0.9436
ROE	60	−61.01	11.86	−6.908	2.158	16.71
GNPA	60	0.07	25.28	12.58	0.683	5.29
NNPA	60	0.02	15.33	5.29	0.399	3.09
GDP	60	−6.60	8.68	3.81	0.708	5.48
IFR	60	3.73	6.70	5.22	0.164	1.27

Table 3 Correlation matrix

		ROA	ROE	GNPA	NNPA	GDP	IFR
ROA	Pearson correlation	1					
	Sig. (2-tailed)						
ROE	Pearson correlation	0.993**	1				
	Sig. (2-tailed)	0.000					
GNPA	Pearson correlation	−0.724**	−0.715**	1			
	Sig. (2-tailed)	0.000	0.000				
NNPA	Pearson correlation	−0.705**	−0.708**	0.882**	1		
	Sig. (2-tailed)	0.000	0.000	0.000			
GDP	Pearson correlation	−0.101	−0.134	0.081	0.191	1	
	Sig. (2-tailed)	0.442	0.308	0.541	0.144		
IFR	Pearson correlation	0.456**	0.458**	−0.406**	−0.555**	0.008	1
	Sig. (2-tailed)	0.000	0.000	0.001	0.000	0.950	

**Correlation is significant at the 0.01 level (2-tailed)

properties, including its range of values, measures of central tendency, and measures of dispersion, are thoroughly summarized in the statistics.

4.2 Correlation Matrix

The correlation matrix is a statistical tool utilized to explore the relationships between variables within a dataset. It provides valuable insights into the strength and direction of associations among different variables. Understanding these correlations is essential for gaining a deeper comprehension of the interdependencies and potential causal connections among the variables being examined.

The correlation matrix table, denoted as IFR, presents six key variables: gross non-performing assets (GNPA), net non-performing assets (NNPA), return on assets (ROA), return on equity (ROE), gross domestic product (GDP), and inflation rate (Table 3).

The correlation matrix table begins by presenting the Pearson correlation coefficient between each variable and itself, which is always equal to 1. The significance of each correlation coefficient is also provided, indicating whether the observed relationship is statistically significant.

The table shows that ROA and ROE have a high degree of relationship, with a strong positive correlation between them of.993. The association between GNPA and NNPA and both ROA and ROE is also negative, demonstrating a connection between higher non-performing asset levels and worse profitability. The GNPA and NNPA have substantial negative associations with ROA and ROE (−0.724 and −0.705 and −0.715 and −0.708, respectively), which are statistically significant at the 0.01 level.

Additionally, the correlations between GDP and the financial variables (ROA, ROE, GNPA, NNPA) are relatively weak, with coefficients ranging from −0.101 to 0.191, suggesting a limited or weak linear relationship. The correlation between IFR and the other variables is also weak, with coefficients ranging from 0.008 to 0.456.

The significance values (Sig. or p-values) indicate the probability of observing the given correlation coefficients by chance. In this table, most of the correlations are statistically significant at the 0.01 level, denoted by "**," indicating a high level of confidence in the observed associations.

Overall, this correlation matrix table provides a comprehensive overview of the relationships between the variables under investigation, offering valuable insights into the interconnectedness and potential influence of these variables on each other.

4.3 *Regression Analysis*

The study's independent variables are gross non-performing assets (GNPA), net non-performing assets (NNPA), the gross domestic product (GDP), and the inflation rate, while its dependent variables are return on assets (ROA) and return on equity (ROE).

Model estimation as return on assets (ROA) as dependent variables

Using the independent variables GDP, inflation rate, gross non-performing assets (GNPA), and net non-performing assets (NNPA), the regression model for forecasting return on assets (ROA) is summarized in the table below. The table provides details on the model fit, the coefficients related to each independent variable, and the analysis of variance (ANOVA).

Model Summary

Model	R	R Square	Adjusted R square	Std. error of the estimate
1	.750[a]	0.562	0.530	0.64690

[a]Predictors: (Constant), IFR, GDP, GNPA, NNPA

ANOVA

Model	Sum of Squares	df	Mean Square	F	Sig.
Regression	29.524	4	7.381	17.638	.000[a]
Residual	23.016	55	0.418		
Total	52.540	59			

[a]Dependent variable: ROA

Coefficients

Model		Unstandardized coefficients		Standardized coefficients		
		B	Std. Error	Beta	t	Sig.
1	(constant)	0.444	0.540		0.823	0.414
	GNPA	−0.095	0.035	−0.535	−2.691	0.009
	NNPA	−0.041	0.068	−0.136	−0.606	0.547
	GDP	−0.006	0.016	−0.033	−0.355	0.724
	IFR	0.121	0.083	0.164	1.468	0.148

[a]Dependent Variable: ROA
Y1 (ROA) = 0.444–0.095 (GNPA) -0.041 (NNPA) -0.006 (GDP) + 0.121 (IFR)...i
[b]Predictors: (Constant), Inflation Rate, GDP, GNPA, NNPA

The dependent variable (ROA)'s variation can be explained by the predictors in the model by roughly 56.2%, according to the coefficient of determination (R-squared) of 0.562. The standardized coefficients (Beta) give an indication of the predictors' relative weight. In comparison to the other predictors, GNPA has the biggest standardized coefficient (−0.535), indicating a substantially higher impact on ROA.

The statistical significance of each coefficient is revealed by the t-statistics and corresponding p-values (Sig.). Statistical significance is commonly defined as a p-value 0.05. The only statistically significant predictor in this situation is GNPA ($p = 0.009$).

In summary, the model suggests that GNPA is a significant predictor of ROA, while NNPA, GDP, and inflation rate do not have a significant relationship with ROA in the given model. However, the overall model is statistically significant and explains a moderate amount of the variance in ROA.

Model estimation as return on equity (ROE) as dependent variables

Model Summary

Model	R	R square	Adjusted R square	Std. error of the estimate
2	.746[a]	0.557	0.525	11.52507

[a]Predictors: (Constant), IFR, GDP, GNPA, NNPA

ANOVA[a]

Model		Sum of squares	df	Mean square	F	Sig.
1	Regression	9184.747	455	2296.187	17.287	.000[b]
	Residual	7305.499	55	132.827		
	Total	16490.247	59			

[a]Dependent variable: ROE
[b]Predictors: (Constant), IFR, GDP, GNPA, NNPA

Coefficients

	Unstandardized coefficients		Standardized coefficients		
Model	B	Std. Error	Beta	t	Sig.
(constant)	7.024	9.618		0.730	0.468
GNPA	−1.569	0.632	−0.497	−2.483	0.016
NNPA	−0.897	1.218	−0.166	−0.736	0.465
GDP	−0.193	0.289	−0.063	−0.669	0.506
IFR	2.162	1.472	0.165	1.468	0.148

[a]Dependent variable: ROE
Y2 (ROE) = 7.024 − 1.569 (GNPA) − 0.897 (NNPA) − 0.193 (GDP) + 2.162 (IFR) + μ

The coefficient of determination (R-squared) is 0.557, indicating that approximately 55.7% of the variance in ROE can be explained by the predictors in the model. The coefficients for NNPA, GDP, and inflation rate are not statistically significant. This means that these predictors do not have a significant relationship with ROE in the given model.

A measurement of the relative importance of the predictors is provided by the standardized coefficients (Beta). GNPA has the biggest standardized coefficient (−0.497), indicating a substantially higher impact on ROE compared to the other factors.

5 Conclusions and Future Research

The relationship between non-performing assets (NPAs) and profitability in Indian public sector banks is thoroughly examined in this paper. The results show that NPAs and profitability metrics, notably return on assets (ROA) and return on equity (ROE), have a significant negative association. According to the data, the selected Indian public sector banks have worse profitability when there are larger numbers of NPAs. Profitability is significantly harmed by bank-specific traits, particularly gross non-performing assets (GNPA). However, net non-performing assets (NNPA) do not significantly impact profitability. It is clear from this that GNPA levels must be under control and reduced in order to increase profitability in Indian public sector banks.

Economically speaking, the study finds that, in the case of the 12 Indian public sector banks under consideration, neither the GDP growth rate nor the inflation rate have a significant impact on how NPAs and profitability are related. These findings suggest that the effects of NPAs on profitability are influenced by a variety of factors, including those other than GDP growth and inflation.

Future research in this field can broaden the scope by considering additional factors that may affect the NPA-profitability relationship. For instance, exploring the impact of regulatory policies, loan recovery mechanisms, and bank-specific strategies could provide a deeper understanding of how Indian public sector banks can effectively manage NPAs and enhance profitability. Furthermore, conducting a comparative analysis between public and private sector banks or investigating the impact of NPAs on other financial performance metrics could yield valuable insights for policymakers and bank management.

References

1. Berger, A. (1995). The relationship between capital and earnings in banking. *Journal of Money, Credit and Banking, 27*(2), 432–456.
2. Bourke, P. (1989). Concentration and other determinants of bank profitability in Europe, North America and Australia. *Journal of Banking & Finance, 13*(1), 65–79.
3. Berger, A. N., & DeYoung, R. (1997). Problem loans and cost efficiency in commercial banks. *Journal of Banking & Finance, 21*(6), 849–870.
4. Demirguc-Kunt, A., & Huizinga, H. (1999). Determinants of commercial bank interest margins and profitability: Some international evidence. *World Bank Economic Review, 13*(2), 379–408.
5. Al-Jafari, M. K., & Alchami, M. (2014). Determinants of bank profitability: Evidence from Syria. *Journal of Applied Finance & Banking, 4*(1), 17–45.
6. Cucinelli, D. (2015). The impact of non-performing loans on bank lending behavior: Evidence from the Italian banking sector. *Eurasian Journal of Business and Economics, 8*(16), 59–71.
7. Duraj, B., & Moci, E. (2015). Factors influencing the bank profitability: Empirical evidence from Albania. *Asian Economic and Financial Review, 5*(3), 483–494.
8. Hashem, H. Y. (2016). Determinants of Egyptian banking sector profitability: Time-series analysis from 2004–2014. *International Journal of Business, Economics and Science Applied Research, 9*(2), 73–78.
9. Al-Homaidi, E. A., Tabash, M. I., Farhan, N. H. S., & Almaqtari, F. A. (2018). Bank-specific and macro-economic determinants of profitability of Indian commercial banks: A panel data approach. *Cogent Economics & Finance, 6*(1), 1–26.
10. Bawa, J. K., Goyal, V., Mitra, S. K., & Basu, S. (2019). An analysis of NPAs of Indian banks: Using a comprehensive framework of 31 financial ratios. *IIMB Management Review, 31*(1), 51–62.

Hands-Free Technology: Acceptance Model for Audiobooks

C. Nagadeepa, R. Vijayakumar, K. K. Jagadeesh, and M. Biju

Abstract Educational content can now be produced and duplicated in text, video, and audio formats thanks to digital technologies. Today's youth actively consume audio content. Audiobooks are frequently requested as a format for instructional content. This chapter's goal is to identify and discuss reading habits and technology-enhanced hands-free and ecofriendly audiobooks. The research examines society's technology acceptance. People are always trying and adopting new technologies. There were 225 respondents who participated in the survey and who gave valuable responses. They were asked to give their opinion before and after using the audiobook for the study purpose. Further, they communicated their acceptance of the new hand-free technology on the basis of its perceived ease of use, its usefulness, and their attitudes toward this use of technology. The findings of the research revealed that respondents significantly reported perceived usefulness and ease of use. Further, they expect more audiobooks for academic purposes and in their reginal languages.

Keywords Audiobook · Hands-free technology · Technology acceptance

1 Introduction

Almost everyone needs to read and improve their literacy. Both respondents and teachers need to work together to improve literacy as part of the education curriculum because it is crucial for developing comprehension and fluency. Experts in studying preparation agree that we need to discover new ways to study [1]. Research in linguistic faculties has discovered students' preferences and the contexts wherein college respondents examine the effects of study habits [2]. Educators should also aim to develop safe surroundings and give college respondents choices to improve study habits. However, different elements can affect college respondents' studying. While there are numerous theories suggesting which motives hamper studying, how

C. Nagadeepa (✉) · R. Vijayakumar · K. K. Jagadeesh · M. Biju
Department of Commerce, Kristu Jayanti College (Autonomous), Bangalore, India
e-mail: vijayakumar@kristujayanti.com; jagadish@kristujayanti.com

J. M. K. P. et al. (eds.), *Business Resilience and Digital Technology in the Post-Pandemic Era*, Contributions to Environmental Sciences & Innovative Business Technology, https://doi.org/10.1007/978-3-031-48075-1_19

much time they lose from only impartial studying, and the effects of overpreparation on studying skills, they have yet to settle on which motivations are best, and engagement is suffering. A 2007 study looking at the National Endowment for the Arts discovered that households were spending much less on books than they had in any other year in the past 20 years, so college respondents won't be able to study texts as much at home. The National Endowment for the Arts (2007) additionally discovered that once readers engage with books and related media in addition to television and video games, they engage more in studying. As a result, college respondents may also not enjoy studying. Many college respondents may be better at impartial studying may during college hours. As a result, it has come to be the obligation of instructors and librarians to at least coordinate and encourage better studying habits for college respondents. College respondents now have better access to digital sources than ever before. Because faculties are dynamic environments that expand as their members expand, assessing school room environments should certainly appeal to education practitioners. Lamb [3] identified the need for instructional reform by conducting practitioner studies. In sports education, in particular in tutorial practices, motion studies are favorable because they enhance one's skills, and conventional studies have provided statistics that are generalizable to other settings [4].

1.1 Audiobooks: Mini Bytes of its History

An audiobook is a book that has been recorded and made available online or on a CD: It can be downloaded from a library computer onto a phone or other listening device. There have been audiobooks in some shape or form for a long time. They existed in the 1930s as chapters or sections of books that were recorded for people with visual impairments on CDs and vinyl records. Two decades later, Long Play (LP) records made it possible to record lengthier pieces, such as complete poems and short plays. An early producer of audiobooks was Caedmon Records, a HarperCollins Publishers brand. Books on Tape (BoT), an audiobook publishing imprint of Random House that emphasizes unabridged audiobook recordings for schools and libraries, was introduced. Books on Tape was previously an independent California-based company before its acquisition by Random House in 2001. Additionally, in the 1950s, the US couple Luce Klein and Arthur Klein established Spoken Arts, which recorded hundreds of literary works by W.B. Yeats, Arthur Miller, and other authors on LP albums.

Naturally, after the invention of the cassette, novels started to be recorded on audiotapes. There were already a number of audiobook businesses by the 1980s, and in the 1990s, the Audio Publishers Association estimated that the market for audiobooks was worth USD 1.5 billion. The Internet opened up a world of possibilities, and in 1997, the podcast and audiobook service Audible introduced the Audible player, a device that could store up to 2 h of audio. After another 25 years, we have arrived on a stage where anyone can listen to any number of books with hours of recorded narration. Contrary to the common misconception that audiobooks further hinder people from developing a reading habit, they have helped people learn their native language, access new libraries for those who are visually impaired, and more.

1.2 Significance of the Study

However, on a larger scale, this has the potential to have an impact on historically underserved student groups. Children nowadays who do not read at a young age may also enter school without the literacy skills that are essential for success in all subject areas. Respondents from disadvantaged homes and the schools that cater to them can also find information from this research that is helpful in restoring missed early literacy skills through the use of audiobooks.

Additionally, this study can be helpful to respondents to inculcate a reading habit for various books, such as not only stories but also different languages. Further, this study can influence respondents to adopt new technologies so that they can increase their vocabulary and pronunciation. This can trigger educators to focus on not only the video content but also new technology, namely audiobooks, because they are all made easy for users to download and use. Studying the effects of audiobooks on pupils while they read was the goal of this investigation. The study determined that the perception of reading books was good after the data had been collected and analyzed, and the analysis showed that reading books aloud can make it easier to read.

2 Literature Survey

The majority of respondents did well on the survey, where false positives were found mostly when a child encountered a word that was very challenging. Additionally, these false positives were less frequent than anticipated.

Helen [5] says that when doing other things, like traveling or cleaning the house, one can listen to an audiobook. Publishers, authors, and journalists seem to be discussing audiobooks primarily in terms of their practicality of consumption and the accessibility they provide by eliminating the need to decipher a book's words on a page. Many of these individuals have written fervent defenses of audiobooks as a medium. The truth remains that when a person is not constrained by demands on their attention, reading abilities, or time by interacting with a physical book, their reading repertory can be expanded [6].

Perrin [7] proved that among younger readers, audio is becoming more and more popular. For instance, 23% of people between the ages of 18 and 29 have listened to an audiobook in the last year, whereas only 8% of adults over the age of 65 have done so.

Johnson and Smith [8] conducted a study to understand the acceptance of audiobooks and explore user attitudes and preferences toward this medium. The researchers collected data through surveys and interviews with audiobook users. The findings revealed that the main factors influencing acceptance were convenience, mobility, and the ability to multitask while listening to audiobooks. Participants also highlighted the importance of narrator performance and voice quality in their enjoyment of audiobooks. The study further explored factors such as perceived value, social influence, and personal preferences in audiobook selection. This research contributes to understanding user perspectives on audiobooks and provides insights for publishers and providers into improving user experiences and promoting the wider acceptance of this medium.

Chen and Lee [9] conducted an empirical investigation to identify the factors influencing the adoption of audiobooks. The study employed a survey questionnaire to collect data from audiobook users. The findings revealed that perceived usefulness, ease of use, and compatibility with users' lifestyles were significant factors impacting the adoption of audiobooks. Participants who found audiobooks convenient, suitable for their busy schedules, and enjoyable as reading experiences were more likely to adopt this medium. Additionally, the study explored the influence of social norms and perceived enjoyment on the intention to adopt audiobooks. The research provides valuable insights into the psychological factors and social influences driving the adoption of audiobooks, which can inform strategies to promote their acceptance among a wider audience.

Thompson and Collins [10] conducted a study to investigate the impact of audiobooks on reading motivation and comprehension skills among students. The research found that incorporating audiobooks into reading activities positively influenced students' motivation to read and enhanced their comprehension skills. Their study suggests that audiobooks can be a valuable tool for promoting reading engagement, particularly for struggling readers or individuals with learning difficulties. The findings highlight the potential benefits of audiobooks as supplementary resources in educational settings.

Lee and Lee [11] conducted a study to examine the effects of audiobooks on vocabulary acquisition in second-language learning. The research involved a group of learners who listened to audiobooks and another group that read print texts. The findings revealed that the group exposed to audiobooks showed significantly higher gains in vocabulary acquisition compared to the traditional reading group. Lee and Lee's study suggests that audiobooks can be beneficial tools for language learners, providing audio support and enhancing their exposure to authentic language use.

Mitchell and Jones [12] conducted a study exploring the role of narrator voice in audiobook enjoyment and immersion. Through surveys and interviews with audiobook listeners, their research examined the impact of different narrator voices on users' overall experiences. The findings revealed that the narrator's voice quality, tone, and ability to convey emotions significantly influenced listeners' enjoyment and immersion in the audiobook. The study highlights the importance of narrator selection and performance in creating a captivating and engaging audiobook experience for the audience.

In a paper titled "Acceptance of Audiobooks: A Review of Literature," Smith et al. [13] provide a comprehensive review of the factors influencing the acceptance of audiobooks as a medium for consuming literature. The authors delve into the growing popularity of audiobooks and the technological advancements that have contributed to their accessibility. By systematically examining existing studies, the review identifies several key factors that influence users' attitudes and behaviors toward audiobooks. These factors include usability and ease of use, convenience, perceived benefits (such as multitasking and enhanced comprehension), perceived drawbacks (such as a loss of control over reading pace), individual differences (such as personality traits and reading habits), and social influence. The review highlights the importance of understanding these factors in order to enhance the acceptance and adoption of audiobooks. It also provides valuable insights for researchers, practitioners, and content creators into the field of audiobook production and distribution. Overall, the review deepens our understanding of the acceptance of audiobooks and its implications for the evolving landscape of literature consumption.

2.1 Benefits of Audiobooks

The simplest method of learning is by reading. Since the vast majority of instructional materials are print or digital texts, reading-challenged kids are severely disadvantaged.

When their learning resources are predominantly text based, respondents with visual weaknesses, attention-span problems, Attention-Deficit / Hyperactivity Disorder (ADHD), or dyslexia confront even-greater obstacles.

Education professionals who work with students who struggle with learning can benefit from using audiobooks. Even pupils who do not have learning disabilities can benefit from the audiobooks' reinforcement of the text's messages to deepen their reading comprehension. The following are four of the benefits that audiobooks offer to enhance learning outcomes both inside and outside of the classroom:

1. Listening to audiobooks makes it easier to learn new words and vocabulary.
2. They provide a double stimulus for learning, in the form of reinforcement.
3. Audiobooks assist respondents who struggle with remembering.
4. They provide support techniques for students who experience attention, dyslexia, and visual impairment problems.

3 Research Methodology

This empirical research was conducted through interviews and survey-based questions on how audiobooks have been accepted by the respondents. Their responses were collected to analyze their perception toward their intention to use audiobooks.

3.1 Research Objectives

1. To determine the acceptance of the new technology—namely audiobooks.
2. To identify the demographical variances among audiobook users.

3.2 Sample Size

For this study, a total of 225 respondents from different regions of India participated. The respondents represented diverse age groups and had varied purposes for using audiobooks, including educational, entertainment, and research purposes.

3.3 Data Collection

Both personal and online interviews were conducted with the respondents. Further, Google forms were used to conduct the data analysis, and structured questionnaires were used and collected for their data. According to the data collected through questionnaires, several hypotheses were formulated and tested.

3.4 Hypotheses

Research hypotheses were developed on the basis of the study objectives. The following hypotheses were framed to determine the respondents' acceptance of audiobooks:

- H1: The quality of an audiobook has a significant influence on respondents' intention to hear/use that audiobook.
- H2: The perceived usefulness of an audiobook has a significant influence on respondents' intention to hear/use that audiobook.
- H3: The ease of use of an audiobook has a significant influence on respondents' intention to hear/use that audiobook.
- H4: The social influence of an audiobook has a significant influence on respondents' intention to hear/use that audiobook.
- H5: The hedonic motivation toward using an audiobook has a significant influence on respondents' intention to hear/use that audiobook.

3.5 Conceptual Framework

The conceptual framework of the study was developed on the basis of the preceding hypotheses. Five independent variables and one dependent variable were identified and formulated for the conceptual framework. Fig. 1 depicts the proposed conceptual framework:

4 Data Analysis

4.1 Demographic Characteristics of Respondents

The first part of the data analysis determined the demographic characteristics of the students who participated and gave their responses to audiobook usage.

The demographic characteristics of the respondents appear in Table 1.

The above table lays out the demographic profiles of the students. Most of them were female (51%) and were in the age group of 26–40 years. According to the data, 25% of audiobooks were used for playing and education purposes (for hearing and academic purposes), and 25% were used for entertainment purposes (for reading stories). The following table shows what the respondents experienced before and after using the audiobooks. For this purpose, a comparative *t*-test was used, and the results appear in Table 2.

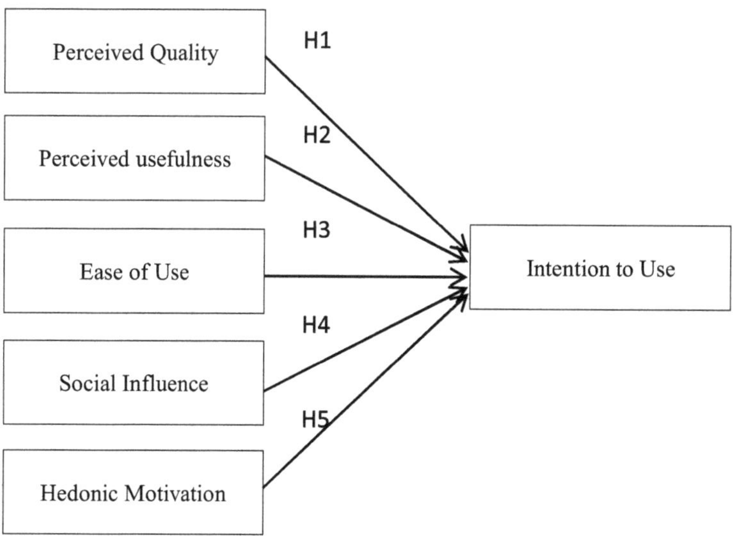

Fig. 1 Conceptual framework of the intention to use an audiobook

Table 1 Demographic characteristics of the respondents

Category	Demographic	Percentage
Gender	Male	48
	Female	51
Age	<12	16
	13–18	17
	19–25	15
	26–40	35
	41–50	12
	>50	5
Purpose	Education purpose—Academic books	25
	Research purpose—Academic books	15
	Entertainment purpose—Story books	25
	Spiritual and devotion—Devotional books	14
	General knowledge—GK books	12
	Miscellaneous	9
Frequency	Daily	55
	Weekly	20
	Monthly	25

Table 2 Before and after using audiobooks: respondents' perceived experiences

Variable	Mean	Standard deviation	Mean difference	Standard deviation	Standard error
Perceived fear	0.84	1.09	0.34	1.06	0.09
Perceived scar	2.54	0.54	−0.08	0.61	0.05
Perceived upset	2.79	0.38	−0.02	0.4	0.04
Perceived nervousness	2.5	0.67	−0.20	0.66	0.06
Perceived happiness	1.89	0.66	0.19	0.67	0.06
Perceived relaxation	1.84	0.7	0.17	0.76	0.07

Variable	t(degree of freedom)	*p*-value	95% confidence interval of mean difference	Cohen's d
Perceived fear	3.61 (130)	< 0.001	0.15 to 0.52	0.32
Perceived scar	−1.58 (130)	0.12	−0.19 to 0.02	0.14
Perceived upset	−0.65 (130)	0.52	−0.09 to 0.05	0.06
Perceived nervousness	−3.44 (130)	0.001	−0.31 to −0.08	0.3
Perceived happiness	3.15 (127)	0.002	0.07 to 0.31	0.28
Perceived relaxation	2.54 (130)	0.01	0.04 to 0.30	0.22

SD Standard Deviation, *SE* Standard Error, t(degree of freedom), *CI* confidence interval

4.2 *Impulse-Buying Behavior of Renewed Products*

The proposed measurement model was assessed by using survey data obtained from participants. To validate the formulated hypotheses, structural equation modeling (SEM) was carried out with the aid of AMOS software. The initial stage of Structural

Table 3 Regression weights of estimated model

Hypothesis	Independent variable	β	Sig (*p*-value)	Result
H_1	Perceived quality	0.055	0.041	Accepted
H_2	Perceived usefulness	0.066	0.011	Accepted
H_3	Ease of use	0.175	0.012	Accepted
H_4	Social influence	0.022	0.035	Accepted
H_5	Hedonic motivation	0.019	0.026	Accepted

Dependent variable: intention to use audiobooks

equation modeling involves evaluating model fitness by comparing diverse model-fit indicators.

The statistical results show the mean of the perceived anxiety of the respondents before and after the audiobook intervention. The pre–post mean differences in perceived fear and in the states of nervousness, happiness, and relaxation were statistically significant in that the *p*-value was less than 0.05. On average, participants felt less fearful and nervous as well as more relaxed and happier after having the experience of listening to an audiobook.

The acceptance values for audiobooks—a new, hands-free technology—were analyzed by using the Technology Acceptance Model (TAM) model, which was developed by Davis in 1989. This model factors in a set of variables to analyze the acceptance values of a technology, which consists of the perceived quality of the technology, the usefulness of the technology, how easy/comfortable it is to use, the social acceptability and influence of the technology, and finally students' hedonic motivation toward using the technology.

Table 3 shows the regression weights of the independent variables on the dependent variable. A regression analysis was conducted to test this study's hypotheses and to measure the respondents' intentions to use/hear audiobooks.

5 Hypotheses: Results on Product-Related Factors

The various influences on whether respondents use audiobooks are discussed in the following subsections.

5.1 Perceived Quality and Intention to Use Audiobooks

The perceived quality of an audiobook had a significant influence on users' intention to use one, at a *p*-value of 0.041, which is less than the significance level of 0.05. This implies that the intention to utilize this technology is significantly influenced by perceptions of the audiobook's quality. The majority of audiobooks offered on numerous websites are excellent, which encourages respondents to use them repeatedly. H1 is therefore acceptable.

5.2 Perceived Usefulness of Audiobooks and Respondents' Intention to Use Them

This factor had a *p*-value of 0.011, which is less than the significance level of 0.05, so users' intentions to utilize audiobooks are significantly influenced by how highly they evaluate its usefulness. Because the respondents may travel great distances to their universities/college premises and are unable to carry their books with them, these audiobooks are helpful for users in their everyday lives in completing tasks fast and enhancing their productivity. When people work and try to multitask, they can't sit and read books. People started listening and doing their household chores or other work simultaneously. Thus, the intention to utilize this technology is significantly influenced by the user's perception of its utility. H2 is therefore accepted.

5.3 Perceived Ease of Use and Intention to Use Audiobooks

This factor had a *p*-value of 0.012, which is less than 0.05, so the perceived usefulness of the audiobooks is significant. No matter one's age, they may use these audiobooks because they are incredibly simple to use and don't require any special technological skills. The majority of respondents preferred to listen than to read. Additionally, the websites' abundance of options makes it simple for kids to use audiobooks. Therefore, the intention to adopt this technology is significantly influenced by the perceived simplicity of using them. H3 is therefore accepted.

5.4 Social Influence and Intention to Use Audiobooks

This factor had a *p*-value of 0.035, which is less than 0.05, for the use of audiobooks. In this technological era, people think that they should use audiobooks, to prove to society that they are up to date on technology. As a result, the majority of them have started utilizing this new technology. Therefore, the intention to adopt this technology is significantly influenced by social influence. Hence, H4 is accepted.

5.5 Hedonic Motivation and Intention to Use Audiobooks

The significant p-value for hedonic motivation toward using audiobooks was 0.026, which is less than 0.05. People believe that audiobooks are accessible both easily and without interruption, even in this stressful and busy environment. As a result, when they utilize this new technology, the majority of them feel calm. So, the desire to employ this technology is significantly influenced by hedonic motivation. H5 is therefore accepted.

6 Conclusion

Despite the widespread notion that listening is less advantageous than reading a physical book, audiobooks have grown in popularity in recent years in the book industry. In particular, the COVID-19 pandemic spurred a lot of changes in various industries. Marketers used technology to reach consumers and persuade them to listen to audiobooks in order to multitask and be more productive. While traveling or doing household chores, people started using audiobooks to increase their productivity. There are a variety of story books and game books available in various websites. At the same time, the availability of academic-related audiobooks remains low. Also, some people would be comfortable with using academic audiobooks if the books were in their respective local languages. Marketers can focus on these kinds of preferences to increase the number of audiobook users.

References

1. Zemelman, S. (2012). Best practice: Bringing standards to life in America's classrooms.
2. Daniels, E., & Steres, M. (2011). Examining the effects of a school-wide reading culture on the engagement of middle school respondents. *RMLE Online, 35*, 1–13. https://doi.org/10.1080/19404476.2011.11462085
3. Lamb, A. (2018). Listeners' advisory part 2: Connecting audiobooks with today's readers. *Teacher Librarian, 45*(3), 62–65.
4. Kozlowski, M. (2020, June 20). Audiobook trends and statistics for 2020. *Good e-Reader*. https://goodereader.com/blog/audiobooks/audiobook-trends-and-statistics-for-2020
5. Helen, A. (2016, November 18). Why I love audiobooks: A print publisher's confession. *Publishers Weekly*. https://www.publishersweekly.com/pw/by-topic/columns-and-blogs/soapbox/article/72077-why-i-loveaudiobooks-a-print-publisher-s-confession.html
6. Miller, L. (2010, June 30). 'War and peace' made easy. *Salon*. https://www.salon.com/2010/06/30/audiobooks_2/
7. Perrin, A. (2019, September 25). One-in-five Americans now listen to audiobooks. https://www.pewresearch.org/fact-tank/2018/03/08/nearly-one-in-five-americans-now-listen-toaudiobooks/.
8. Johnson, M., & Smith, L. (2018). Understanding the acceptance of audiobooks: A study of user attitudes and preferences. *Journal of Information Science, 44*(3), 345–361.

9. Chen, H., & Lee, W. (2019). Factors influencing the adoption of audiobooks: An empirical investigation. *Computers in Human Behavior, 93*, 78–89.
10. Thompson, E., & Collins, S. (2020). Exploring the impact of audiobooks on reading motivation and comprehension skills. *Journal of Educational Psychology, 112*(4), 678–693.
11. Lee, J., & Lee, S. (2019). Effects of audiobooks on vocabulary acquisition in second language learning. *Language Learning & Technology, 23*(2), 157–177.
12. Mitchell, R., & Jones, H. (2021). The role of narrator voice in audiobook enjoyment and immersion. *Journal of Media Psychology, 17*(3), 432–449.
13. Smith, A., Johnson, B., & Davis, C. (2021). Acceptance of audiobooks: A review of literature. *Journal of Media Psychology, 15*(2), 178–203.

Antecedents of Customer Experiences: An Investigation into Utilitarian and Hedonic Cues in Quick-Bite Restaurants

T. K. Sateesh Kumar, P. T. Retheesh, and N. M. Vipulkumar

Abstract Focusing on customer experiences is the latest strategy adopted by marketers to build competitive advantage. This study aims to determine the antecedents of such experiences and how people behave with each other in the quick-bite restaurants in Bangalore. It also analyses the correlations between such latent factors, and it features a model for constructing utilitarian cues and hedonic cues of customer experiences. This is a descriptive study, wherein the survey method collects primary data by administering structured Likert-scale tests to 144 respondents. The secondary data are utilised for conceptualising the model, and the primary data are analysed via principal component analysis (PCA) and confirmatory factor analysis (CFA) by using Statistical Package for Social Sciences (SPSS) and Microsoft Excel (MS Excel). The study links the cues and customer experience. Both the hedonic cues and the utilitarian cues are compared to determine how they affect customer experiences, specifically in quick-bite restaurants.

Keywords Customer experience · Utilitarian and hedonic cues · Quick-bite restaurants

1 Introduction

Marketing evolved over time, wherein the products are marketed through better offerings, and nowadays, marketers are trying to develop more-effective strategies. Again, over time, marketers have made in-roads by building desirable customer experiences. This customer experience encompasses marketing, consumer research, retailing, service dominant (SD), service design, online marketing, branding, and experiential marketing [1]. The customer experience (CX) is an internal, subjective response to a company's offerings, perhaps from direct or indirect contact with the company [2].

T. K. Sateesh Kumar (✉) · P. T. Retheesh · N. M. Vipulkumar
Department of Commerce, Kristu Jayanti College (Autonomous), Bangalore, India

J. M. K. P. et al. (eds.), *Business Resilience and Digital Technology in the Post-Pandemic Era*, Contributions to Environmental Sciences & Innovative Business Technology, https://doi.org/10.1007/978-3-031-48075-1_20

This study is restricted to quick-bite restaurants, also called fast-food restaurants, which is one of the fastest growing sectors, at 18%. The definition of the *quick-bite restaurant* is one where the customers are able to consume food and beverages [3] with minimal table service. Thus, this study on quick-bite restaurants aims to understand subjective and objective customer experiences and their cues.

Berry et al. [4] described the entire customer experience. First, CXs derive from the mind of the customer, and second, CXs are emotional. Moreover, the emotional aspect is either mechanic—generated by things—or humanic [5]—generated by people. These humanic emotions take precedence over reasoning in customer decision-making.

Customer experience is informed by eight key factors: service marketing, consumer research, retailing, service dominant (SD) logic, service design, online marketing, branding, and experiential marketing. Various models have been developed to understand the customer experience: the SERVQUAL model [6], the eTailQ experience [7], the E-S-QUAL model [8], the consumption experience model, the customer experience model [9], the conceptual model of customer experience [10], and the EXQ framework [11]. The researcher conducted the study on several industries, including banking, online retail, aviation, telecommunication, and media.

The theoretical approaches to customer experiences are overwhelming; however, CXs are confined to the creation of experience, its measurement, and mapping the outcomes of such CXs. The main pre-requisite for understanding CXs is to understand the role of customer stimuli and cues [12].

In the process of mastering the managing of such customer experiences, many business leaders have tried to build satisfying customer experiences. Thus, they have poured their expertise into learning the clues and buying processes of customers that culminates in the practice of creating value throughout the customer journey.

In order to influence the customer, many entrepreneurs have striven to understand the facets of offering the best products and services. These facets are called cues. In theory, they are stimuli that can influence customer behaviour and experiences.

Numerous cues together determine the customer experience. These may be summarised as follows: visual cues, tactile cues, social cues, and emotional cues. Visual cues include the appearance and design of the product and services, such as the packaging, colours, logos, and shape of the product; tactile cues that use touch or movement to help people understand things; social cues are the interactions between customers, staff members, and influencers, including influence by means of via the recommendations of others, online reviews, and celebrity endorsements; and emotional cues include the feelings associated with the product or services.

These cues can enhance the customer experience by having visual, auditory (sound), olfactory (scent), tactile (touch and feel), and social effects. But how such cues are interpreted, such as whether they are utilitarian or hedonic, depends on in the customer's perspective. Utilitarian cues are those data or stimuli that derive from product quality, price, etc., which directly reflect the functions and usefulness of the products. In contrast, hedonic cues are those that instigates emotional and sensory cues to elicit emotional and pleasurable experiences in the customer.

The majority of the studies in the literature on these cues seldom report on how these cues interact and correlations among them. This paves way for the first research question (RQ1): What are the indicators that build constructs (i.e. cues) and how are the relationship among them? The second research question (RQ2): Is these cues are applicable to the customer experience (constructs) in quick bite restaurants. Thus, the following three objectives derived from the research questions, reviews, and this gap in the literature:

1. To identify the antecedents and constituents that build customer experiences in quick-bite restaurants
2. To analyse the correlations between customer experiences and their utilitarian cues and hedonic cues (RQ1)
3. To validate the constructs and develop a model for the utilitarian and hedonic cues of customer experiences in quick-bite restaurants (RQ2)

2 Methodology

This study aims to identify the facets of customer experience that are built through various cues in quick-bite restaurants; it is therefore a descriptive study, wherein the primary data are collected by using the survey method. This research also utilises secondary data. The primary data are collected through 15 structured five-point Likert-scale tests, administered to and collected from 144 respondents, consisting of the customers of quick-bite restaurants in Bangalore. The selected quick-bite restaurants are the top five in India: Dominos, McDonald's, KFC, Pizza Hut, and Subway, according to sales volume and the number of outlets.

Statistical tools—principal component analysis (PCA) and confirmatory factor analysis (CFA) in Statistical Package for Social Sciences (SPSS) and Microsoft Excel—were utilised extensively.

The cues were considered as premises for developing questionnaires according to Javed et al. [13], and they measure the perception of the customer experiences [12]. Perpetual cues are based on data from humans' five senses. The objectives resulted in the development of hypotheses and a conceptual model. The first objective was to understand the theoretical background; as discussed in the review of the literature, there are utilitarian cues and hedonic cues. The second major objective was to determine their effects on customer experiences. Figure 1 displays the conceptual model.

The second objective yielded an alternative for analysis: to identify the utilitarian cues and hedonic cues that influence customer experiences. Therefore, H1*a* aims to validate the factors by using principal component analysis and confirmatory factor analysis (H1*a* is based on a framework developed by Shrestha [14] and Tinsley and Tinsley [15]).

$H1_a$: Latent factors influence the measured indicators.

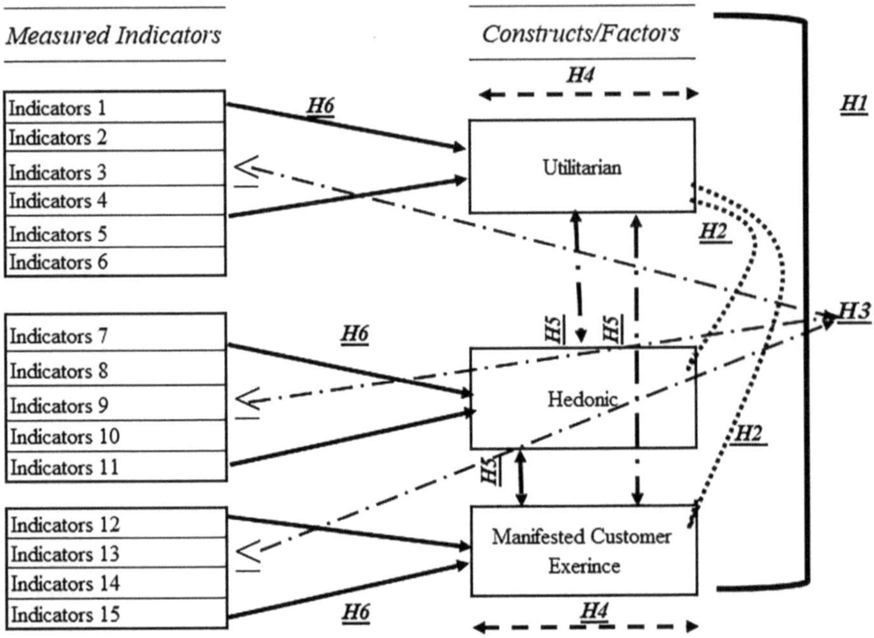

Fig. 1 Conceptual model of cues in customer experiences

Once the factors have been identified, the second phase of the objective is to measure the degree of correlation, which gave rise to $H2_a$ ($H2_a$ is based a framework developed by Shrestha [14] and Kim et al. [16]).

$H2_a$: A significant correlation exists between several latent factors, namely utilitarian cues, hedonic cues, and customer experiences.

These factors are theoretical representations of the manifested constructs. As per objective three, the researcher aimed to validate these factors' influences. The factors undergo validation through convergent tests (by using average variance extracted, aka AVE) and composite reliability and discriminant tests. Hypotheses H3, H4, and H5 were derived from the framework specified by Hair [17, 18].

$H3_a$: There is significant internal consistency among the measured indicators (composite reliability > 0.7).

$H4_a$: There are agreement and convergence among the indicators (AVE > 0.5).

Another way to measure the strength of an influence is how well the loaded factors differ from the other factors: how distinct they are from one another. In order to test a construct according to its distinctiveness, another hypothesis (H5) was developed.

$H5_a$: The measured indicators differentiate between conceptually distinct constructs (discriminant validity).

Finally, to achieve the primary objective of this study, the last hypothesis (H6) was developed. This hypothesis is designed in line with RQ1 and RQ2 and to confirm the conceptual model. Thus, a construct validity analysis was carried out to validate how well the data fit into the confirmation model.

This model is called a measurement model because it measures relationships among latent variables, but not their directional natures (unlike a structural model). The measurement model fits into the theoretical model; in other words, the measured indicator data fit into the theoretical model. The last hypothesis, which was based objective 3, relied on studies conducted by Sunil [19] and Sürücü [20].

$H6_a$: The observed variables or indicators are not measuring the intended latent constructs or factors, and the measurement model does not fit the data well.

3 Results and Analysis

The analysis was carried out starting with the validity of the instruments used for data collection. The four validity measures of the questionnaires are face validity, content validity, criterion validity, and construct validity. Face validity was measured on a five-point Likert scale. Content validity measured the two main types of cues—hedonic cues and utilitarian cues—by using five questionnaires for each. Criterion validity was measured according to correlation data, i.e. how each cue correlated with the others; this is reflected in hypothesises H4 and H3. Finally, the study aimed to determine the construct validity, which was achieved by using various tests, as discussed in the hypothesis-testing section.

For reliability, the Cronbach's alpha test was carried to test the coefficient of variation among the questionnaires. Table 1 shows how significant the coefficient of variation (Cronbach's alpha) is.

The second phase of data analysis was carried out in line with objective 2, wherein the research aimed to identify any latent factors and how strong their influences were. For this purpose, the Kaiser–Meyer–Olkin (KMO) and Bartlett's test were carried. This test determines the sampling's adequacy and the correlations among the various indicators: >0.7 (0.837) and significance level of <0.05 (0.00). Thus, these values show that the requirements for carrying out a factor analysis were fulfilled (*all* measured indicators must be fulfilled) (Table 2).

Furthermore, the research used a principal component analysis (PCA) to determine which factors to include in the factor analysis. This process involves analysing communalities: the degree of variance that each measured indicator shares with underlying latent factors/constructs. Beavers [21] and Eaton [22] reported levels of 0.24 to 0.40 as cutoffs and > 0.70 as ideal.

Table 1 Cronbach's alpha (reliability statistics)

Cronbach's alpha	No. of items
0.878	15

Table 2 KMO and Bartlett's test

The Kaiser-Meyer-Olkin measure of sampling adequacy		0.837
Bartlett's test of sphericity	Approx. Chi-square	980.600
	Df	105
	Significance	0.000

Table 3 Communalities

	Initial	Extraction
Speed of the service	1.000	0.423
Visibility and information availability	1.000	0.598
Quality of products	1.000	0.439
Accuracy of product information	1.000	0.624
Meeting expectation	1.000	0.230
In-store delivery	1.000	0.628
Easy access to catalogue	1.000	0.574
Appearance of product	1.000	0.675
Taste of product	1.000	0.679
Smell of product	1.000	0.658
Texture of product	1.000	0.634
Increased loyalty	1.000	0.688
Word of mouth	1.000	0.760
Swapping of brands	1.000	0.731
Brand image	1.000	0.772

Extraction method: principal component analysis

Table 3 shows the communalities among CXs: word of mouth (0.760), swapping of brands (0.731), and brand image (0.772). However, the following communalities only moderately influenced the hedonic cues: the appearance of product (0.675), taste of product (0.679), smell of product (0.658), and texture of product (0.634).

In the next phase, the factor analysis explains the total variance, as shown in Table 4. There are 15 measured indicators that mirror their corresponding components, where initial eigenvalues >1 are considered as factors for analysis. The eigenvalues are the squared values of all the indicators of an underlying factor. Only those factors whose eigenvalues were > 1 were considered.

Table 4 shows the three factors (which were extracted by using the PCA) that met the criterion of an eigenvalue of >1: component 1–5.641; component 2–1.971; and component 3–1.500. These three components make up the cumulative 60.742%.

After the study extracted these factors, it reaffirmed the three factors because the scree plot produced an elbow curve, as depicted in Fig. 2.

On the X-axis, Fig. 2 shows the number of components (15), and the Y-axis shows the eigenvalues. The coordinates of one eigenvalue include three components.

Nevertheless, without properly distributing the factor loading, the researcher will have difficulty in identifying the underlying the factors. Hence, in the next stage, the research rotated the components matrix, which properly distributed the loading among the factors/components.

Table 4 Total variance

	Initial eigenvalues			Extraction sums of squared loadings			Rotation sums of squared loadings		
Component	Total	% of variance	Cumulative %	Total	% of variance	Cumulative %	Total	% of variance	Cumulative %
1	5.641	37.605	37.605	5.641	37.605	37.605	3.131	20.876	20.876
2	1.971	13.140	50.744	1.971	13.140	50.744	3.105	20.702	41.578
3	1.500	9.998	60.742	1.500	9.998	60.742	2.875	19.164	60.742
4	0.978	6.520	67.263						
5	0.812	5.415	72.678						
6	0.715	4.766	77.444						
7	0.641	4.272	81.716						
8	0.561	3.740	85.456						
9	0.409	2.730	88.185						
10	0.391	2.609	90.794						
11	0.359	2.396	93.190						
12	0.323	2.155	95.345						
13	0.267	1.777	97.122						
14	0.224	1.492	98.615						
15	0.208	1.385	100.000						

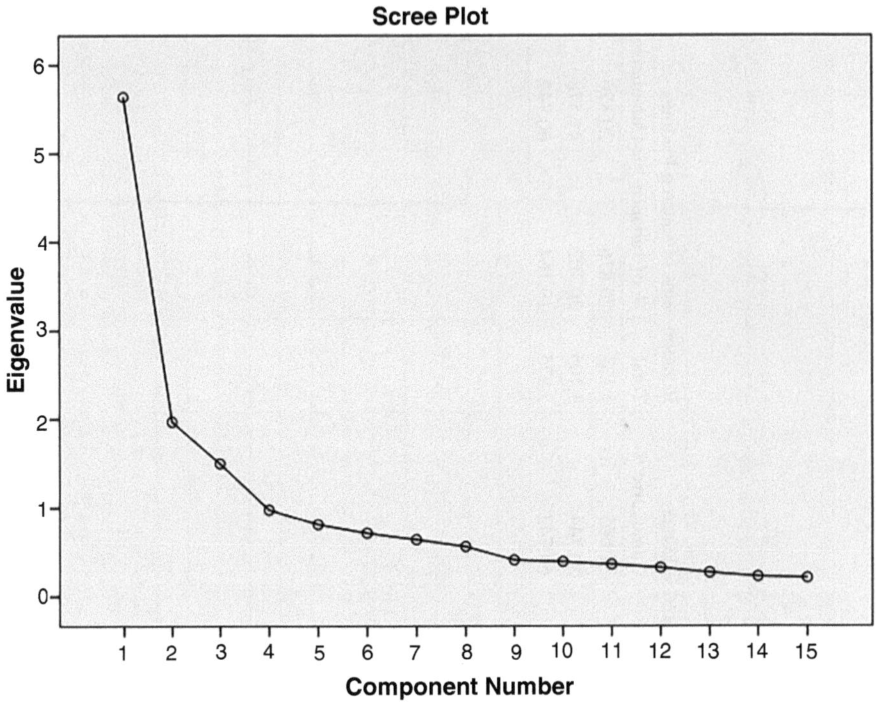

Fig. 2 Scree plot

Table 5 categorises the three major types of factors: the components that represent utilitarian cues (Component 1), hedonic cues (Component 3), and customer experiences (Component 2).

The next analysis covers the transformation of the component, from before to after the component rotation has been completed. It also shows why the extraction method and rotation method were adopted.

Table 6 shows the transformation component matrix before and after the rotation. Component 1 transformed from 0.554 to 0.618; Component 2 from 0.325 to 0.823; and Component 3 from 0.108 to 0.716.

This component plot (Fig. 3) represents the subsequent phase of analysis, wherein the rotated matrix is converted to show how the variables are visually grouped.

Figure 3 shows three components on a dimension plot, where the variables are plotted in proximity to one another:

Table 5 Rotated component matrixa

	Component		
	1	2	3
Speed of the service	0.566	0.172	0.270
Visibility and information availability	0.724	0.167	0.213
Quality of products	0.547	0.213	0.308
Accuracy of product information	0.769	−0.047	0.177
Meeting expectation	0.096	−0.064	0.465
In-store delivery	0.749	0.253	0.047
Easy access to catalogue	0.727	0.154	0.145
Appearance of product	0.171	0.285	0.751
Taste of product	0.196	0.161	0.784
Smell of product	0.262	0.166	0.750
Texture of product	0.173	0.117	0.768
Increased loyalty	0.263	0.776	0.132
Word of mouth	0.169	0.850	0.096
Swapping of brands	0.101	0.836	0.147
Brand image	0.152	0.858	0.109

Extraction method: principal component analysis
Rotation method: varimax with kaiser normalisation
[a]The rotation converged in five iterations

Table 6 Component transformation matrix

Component	1	2	3
1	0.618	0.557	0.554
2	0.325	−0.823	0.466
3	−0.716	0.108	0.690

Extraction method: principal component analysis
Rotation method: varimax with kaiser normalisation

$$\frac{\left(\sum_{i=1}^{i} \lambda_i\right)^2}{\left(\sum_{i=1}^{i} \lambda_i\right)^2 + \left(\sum_{i=1}^{i} 1 - \lambda_i^2\right)} \qquad \text{AVE} = \frac{\sum_{i=1}^{n} L_i^2}{n}$$

where λ represents the factor loading into the construct; L represents the squared factor loading; *i* represents the *i*th item; and *n* represents the number of items in the unobserved latent variable (in this case, the construct), as mentioned in Brown [23].

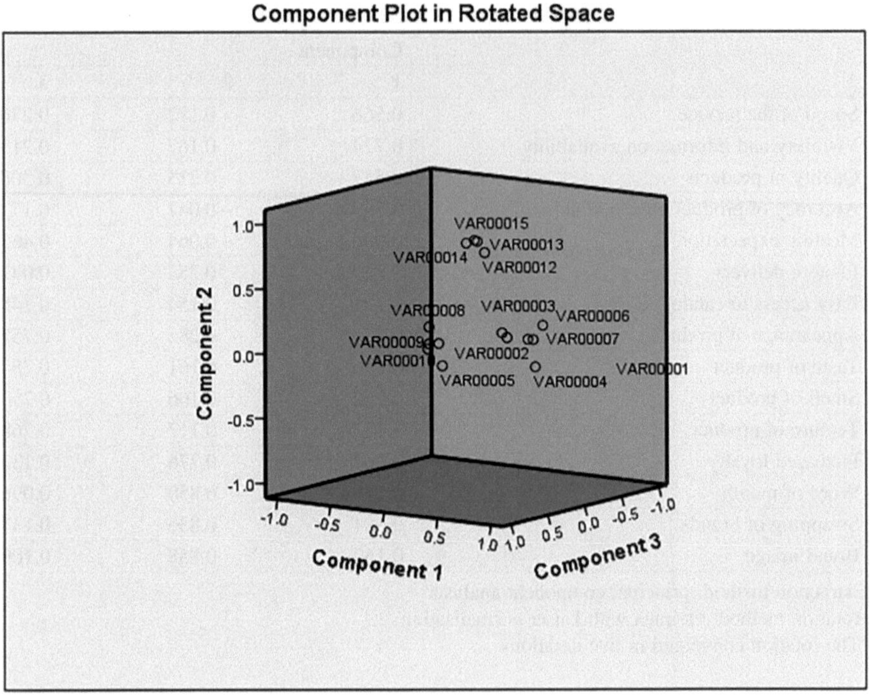

Fig. 3 Component plot in rotated space

4 Composite Reliability

The composite reliability indicates the degree to which the items in the scale consistently measure the same construct (in Table 7, the UC value is 0.75; HC is 0.84; and CX is 0.88). The threshold level for composite reliability is a CR value >0.7 [24, 25]; thus, the construct has achieved the composite reliability, and therefore, $H4_a$ is accepted.

Table 7 The constructs: CRs, AVEs, UCs, HCs, and CXs

	Composite Reliability CRs	AVEs	Utilitarian cues (UCs)	Hedonic cues (HCs)	Customer experiences CXs
Utilitarian cues (UCs)	0.751	0.501	0.708		
Hedonic cues (HCs)	0.845	0.578	0.571	0.760	
Customer experiences (CXs)	0.884	0.655	0.408	0.435	0.809

5 Convergent Validity

According to the data in Table 7, the AVE is >0.5 (threshold level taken from [14, 17, 18]), so the researcher achieved convergent validity. On composite reliability, the standardised regression weights of the measured variables to the constructs are presented in Table 8.

6 Discriminant Validity

In the next phase, the researcher tested the discriminant validity of the construct. This paved for assessing $H5_a$. The discriminant validity of a construct is the degree to which it is empirically distinct from the other constructs in a measurement model. For this reason, the current study adopted the intercorrelation with other constructs. Thus, if the square root of AVE is greater than the interconstructed correlation, then the construct has been validated. In Table 7, for example, the UC (0.708), HC (0.76), and CX (08.09) values are greater than the interconstruct correlation (with other constructs). Therefore, all the constructs were validated for discriminant validity, implying the acceptance of $H5_a$.

Finally, the primary objective was to validate the conceptual model with data that fit with the measurement model. This objective informed $H6_a$, for which a test was carried out to examine whether the measured indicators were properly loaded into the designated factor. This was carried out via a confirmatory factor analysis, which is a specific kind of factor analysis.

This model denotes the relations between the latent constructs and the measured variables. The path analysis is shown in Fig. 4, where three latent factors—namely UCs, HCs, and CXs—reflect the measured indicators (variables) that are affected. More specifically, indicators can be affected by errors (unexplained variance). The standardised model, which is represented by the values ranging from 0 to 1 [26, 27],

Table 8 Standardised regression weights (group 1—default model)

Measured indicators		*Constructs*	*Estimates*
Visibility and information availability	<–	Utilitarian cues	0.709
Accuracy of product information	<–	Utilitarian cues	0.697
Appearance of product	<–	Hedonic cues	0.769
Taste of product	<–	Hedonic cues	0.799
Smell of product	<–	Hedonic cues	0.781
Texture of product	<–	Hedonic cues	0.688
Increased loyalty	<–	Customer experience	0.773
Word of mouth	<–	Customer experience	0.841
Swapping of brands	<–	Customer experience	0.805
Brand image	<–	Customer experience	0.817
Easy access to catalogue	<–	Utilitarian cues	0.717

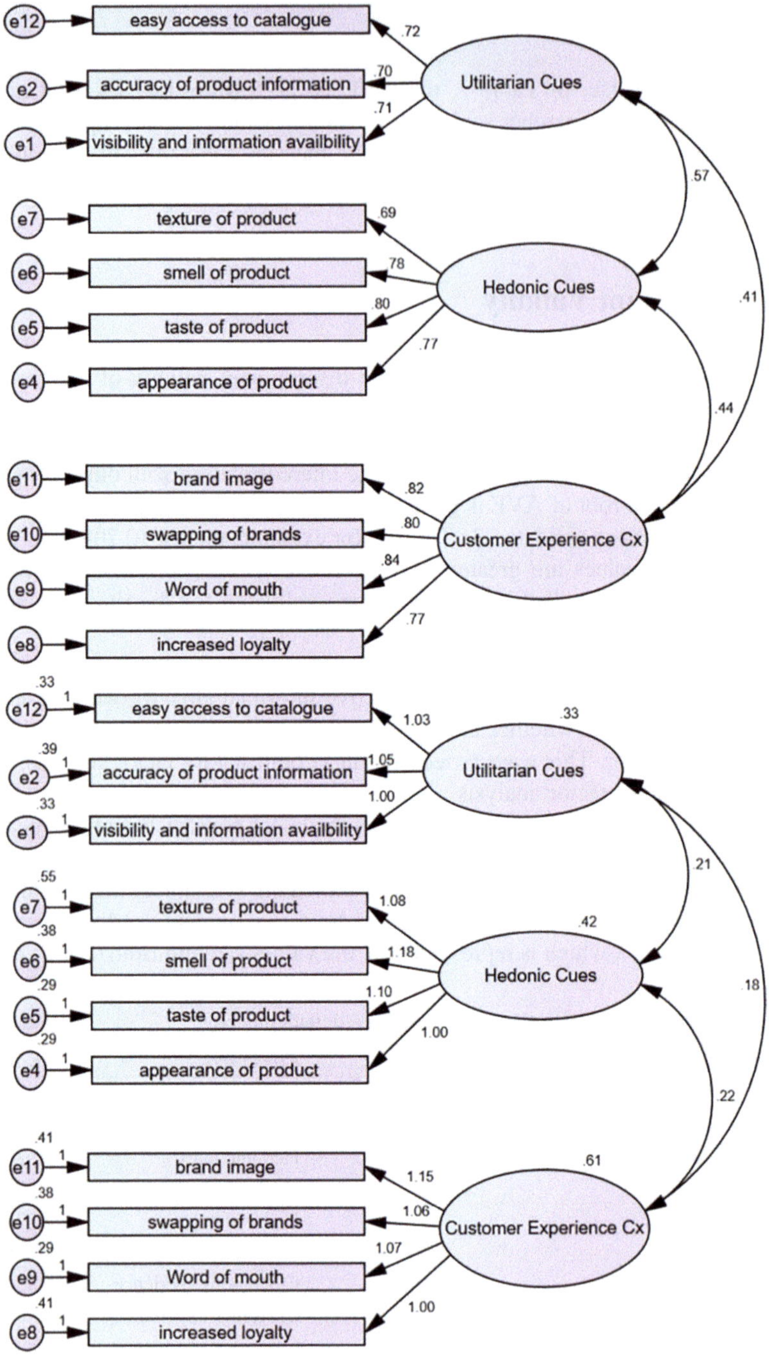

Fig. 4 Confirmatory factor model (standardised (α_r) and unstandardised (α_Σ))

indicates that the model is based on the correlation matrix. Whereas the right side of Fig. 4 depicts the model that is based on the correlation matrix (denoted by α_r), the left side of the model depicts the covariance matrix (denoted by α_Σ). The relations between the variables are indicated by single-headed arrows and double-headed arrows: The single-headed arrows denote regressions, and the double-headed arrows denote correlations. The unexplained variance is denoted by e_i, which is encircled by and attached to the independent variables.

7 Discussion

The validity and the reliability of a construct are the most pivotal goals of the study; more specifically, the scales are at the root of the study's reliability. Hence, the measured indicators from questionnaires were tested for reliability by using Cronbach's alpha, and the researcher was able to obtain significant reliability statistics ([28] > 0.6) for the measured scales (see Table 1; Cronbach's alpha of 0.878). Note that the 15 instruments used for data collection had internal consistency.

The PCA analysis provided three major factors, as shown in Table 4: the constructs of HC, UC, and CX had a significant commonality [14, 24], that of >0.6074. In Fig. 2, which contains a scree plot determining the number of factors, the three major components (factors) correspond to those in Table 4. The analyses require good distributions of measured variables in the components, i.e. the indicators (measured variables) loading into each factor. Thus, Table 5 and 6 and Fig. 3 all feature the same three components: UCs, HCs, and CXs.

H4 is accepted because the constructs of the utilitarian cues (0.751), hedonic cues (0.845), and customer experiences (0.884) were revealed to have significant internal consistency. Using the CR value is preferred [24, 29] over using Cronbach's alpha. Notwithstanding, the Cronbach's alpha of the study was significant (> 0.87), as shown in Table 1.

Convergent reliability measures AVE, more specifically the extent to which different items that are intended to measure the same construct are related to one another. The recommended level for convergent validity is an AVE value >0.5 [30, 31]. However, some indicators were statistically insignificant that were not considered for further analysis, such as the speed of the service, the quality of the products, whether the products meet expectations, and in-store delivery (mainly UC indicators).

In this study, as depicted in Table 2, the latent constructs are able to manifest average variance extracted values >0.5 (UC of 0.501; HC of 0.578; and CX of 0.655). The researcher accepts H4 because both composite reliability and convergent reliability were exhibited in all the constructs. In Table 7, the square root of the AVE value is greater than the interconstruct correlation (0.708; HC 0.76; and CX 0.809), so the measurement model has discriminant validity.

The primary objective of this study was to develop a confirmative model that would collect data to identify the antecedents and constituents that build customer experiences in quick-bite restaurants. This was achieved when the path analysis depicted that each measured indicator was loading into a latent construct, such as CX, UC, or HC.

8 Findings and Suggestions

This study highlighted the cues that determine customer experiences in quick-bite restaurants. The major findings were that the data collected confirmed the following cues: hedonic cues, utilitarian cues, and customer experiences. However, not all of the 15 measured indicators were able to reach significance in the study. Of the measured indicators, 11 were grouped into three factors. This implies that business leaders diversify their product positioning not in a utilitarian war but rather in a hedonic way, which can further boost CXs. On standardisation, in Fig. 4, the path analysis showed significant correlations between the latent constructs. The CX construct had very significant variance with the HC and UC constructs. This suggests that the respondents were more inclined towards HC (0.42) and slightly more inclined towards UC (0.33). There are four more statistically significant indicators that explain HCs and CXs.

These results reaffirm that Indian customers (who were the population sample under study) are more interested in the selected quick-bite restaurants for hedonic reasons rather than for utility ones, as the majority of these restaurants intend [13, 32].

This study was limited to the geographical region of Bangalore, and its measured indicators did not encompass all the various experiences that customers may have. However, future studies can continue this work by exploring auditory cues, because this study did not. Furthermore, future studies can research other CXs with similar industries.

9 Conclusion

The primary objective of this study was to develop a model for and to confirm the validity of underlying latent constructs, such as UCs, HCs, and CXs. This objective was achieved by assessing and confirming the validity of the various constructs under study. This study identified the strength of each construct and depicted its results in tables and figures. In conclusion, the study scientifically established that the UCs, HCs, and CXs of quick-bite restaurants influence one another.

References

1. Becker, L., & Jaakkola, E. (2020). Customer experience: Fundamental premises and implications for research. *Journal of the Academy of Marketing Science, 48*, 630–648.
2. Meyer, C., & Schwager, A. (2007). Understanding customer experience. *Harvard Business Review, 85*(2), 116.
3. Parsa, H. G., Shuster, B. K., & Bujisic, M. (2020). New classification system for the US restaurant industry: Application of utilitarian and hedonic continuum model. *Cornell Hospitality Quarterly, 61*(4), 379–400.
4. Berry, L. L., Carbone, L. P., & Haeckel, S. H. (2002). Managing the total customer experience. *MIT Sloan Management Review*.
5. Carbone, L., & Haeckel, S. (1994). Engineering customer experiences. *Marketing Management, 3*, 8–19.
6. Parasuraman, A. B. L. L., Zeithaml, V. A., & Berry, L. (1988). SERVQUAL: A multiple-item scale for measuring consumer perceptions of service quality, *64*(1), 12–40.
7. Wolfinbarger, M., & Gilly, M. C. (2003). eTailQ: Dimensionalizing, measuring, and predicting etail quality. *Journal of Retailing, 79*(3), 183–198.
8. Parasuraman, A., Zeithaml, V. A., & Malhotra, A. (2005). ES-QUAL: A multiple-item scale for assessing electronic service quality. *Journal of Service Research, 7*(3), 213–233. https://doi.org/10.1177/1094670504271156
9. Gentile, C., Spiller, N., & Noci, G. (2007). How to sustain the customer experience: An overview of experience components that co-create value with the customer. *European Management Journal, 25*(5), 395–410.
10. Verhoef, P. C., Lemon, K. N., Parasuraman, A., Roggeveen, A., Tsiros, M., & Schlesinger, L. A. (2009). Customer experience creation: Determinants, dynamics and management strategies. *Journal of Retailing, 85*(1), 31–41.
11. Klaus. (2012). EXQ: a multiple-item scale for assessing service experience. *Journal of Service Management, 23*(1), 5–33. https://doi.org/10.1108/09564231211208952
12. Havíř, D. A. (2017). Comparison of the approaches to customer experience analysis. *Economics & Business, 31*(1).
13. Javed, S., Rashidin, M. S., Zhu, M., Xu, Z., Jian, W., & Zuo, S. (2021). Combined effects of drivers and impact of customer satisfaction on brand loyalty: The contingent effect of social trust. *SAGE Open, 11*(1), 21582440211003566.
14. Shrestha, N. (2021). Factor analysis as a tool for survey analysis. *American Journal of Applied Mathematics and Statistics, 9*(1), 4–11.
15. Tinsley, H. E. A., & Tinsley, D. J. (1987). Uses of factor analysis in counselling psychology research. *Journal of Counseling Psychology, 34*(4), 414–424. https://doi.org/10.1037/0022-0167.34.4.414
16. Kim, J. J., Lee, J. S., & Han, H. (2023). Tangible and intangible hotel in-room amenities in shaping customer experience and the consequences in the with-corona era. *International Journal of Contemporary Hospitality Management, 35*(2), 657–681.
17. Hair, J. F., Ringle, C. M., & Sarstedt, M. (2011). PLS-SEM: Indeed a silver bullet. *Journal of Marketing Theory and Practice, 19*(2), 139–152.
18. Hair, J. F., Risher, J. J., Sarstedt, M., & Ringle, C. M. (2019). When to use and how to report the results of PLS-SEM. *European Business Review, 31*(1), 2–24.
19. Ahire, S. L., & Devaraj, S. (2001). An empirical comparison of statistical construct validation approaches. *IEEE Transactions on Engineering Management, 48*(3), 319–329.
20. Sürücü, L., & MASLAKÇI, A. (2020). Validity and reliability in quantitative research. *Business & Management Studies: An International Journal, 8*(3), 2694–2726.
21. Beavers, S., Lounsbury, J. W., Richards, J. K., Huck, S. W., Skolits, G. J., & Esquivel, S. L. (2013). Practical considerations for using exploratory factor analysis in educational research. *Practical Assessment, Research and Evaluation, 18*, 1.

22. Eaton, P., Frank, B., Johnson, K., & Willoughby, S. (2019). Comparing exploratory factor models of the brief electricity and magnetism assessment and the conceptual survey of electricity and magnetism. *Physical Review Physics Education Research, 15*(2), 020133.
23. Brown, T. A., & Moore, M. T. (2012). Confirmatory factor analysis. *Handbook of structural equation modeling, 361*, 379.
24. Hair, J. F., Black, W. C., Babin, B. J., & Anderson, R. E. (2014). *Multivariate data analysis* (7th ed.). Pearson.
25. Raykov, T. (1997). Estimation of composite reliability for congeneric measures. *Applied Psychological Measurement, 21*(2), 173–184. https://doi.org/10.1177/01466216970212006
26. Falk, C. F., & Savalei, V. (2011). The relationship between unstandardized and standardized alpha, true reliability, and the underlying measurement model. *Journal of Personality Assessment, 93*(5), 445–453.
27. Schreiber, J. B., Nora, A., Stage, F. K., Barlow, E. A., & King, J. (2006). Reporting structural equation modeling and confirmatory factor analysis results: A review. *The Journal of Educational Research, 99*(6), 323–338.
28. Nunnally, J. C., & Bernstein, I. H. (1994). *Psychometric theory* (3rd ed.). McGraw-Hill.
29. McDonald, R. P. (1999). *Test theory: A unified treatment*. Lawrence Erlbaum Associates.
30. Fornell, C., & Larcker, D. F. (1981). Evaluating structural equation models with unobservable variables and measurement error. *Journal of Marketing Research, 18*(1), 39–50. https://doi.org/10.1177/002224378101800104
31. Gao, Y., Li, X., & Sun, X. (2020). Influence of cross-cultural adjustment on expatriate job performance in China: The mediating roles of job satisfaction and job commitment. *Asia Pacific Journal of Management, 37*(2), 591–611. https://doi.org/10.1007/s10490-019-09647-6
32. Nandini, A. S. (2014). McDonald's success story in India. *Journal of Contemporary Research in Management, 9*(3), 21.

Provision of IoT in Privileged and Sectors in India

Arti Singh, V. Santhosh, and Kenneth Wilson Bavachan

Abstract The provision of Internet of things (IoT) technology in India has been uneven, with privileged sectors such as urban areas and high-income communities receiving greater access and benefits compared to unprivileged sectors such as rural areas and low-income communities. Studies have shown that the main barriers to IoT adoption in unprivileged sectors in India include lack of infrastructure, low levels of digital literacy, and limited access to affordable devices and services. Additionally, there are also cultural and language barriers that further hinder the provision of IoT in these communities. However, there have been efforts by the government and private organizations to bridge the digital divide and provide IoT solutions to unprivileged sectors. For example, the government has launched programs to promote digital literacy and provide Internet connectivity to rural areas. Private companies have also initiated projects to develop low-cost IoT devices and provide IoT-based services in areas such as agriculture and healthcare. The present chapter purports to study about the scope of IoT in privileged and unprivileged sectors, along with the data protection provisions for consumer IoT, its uses, relevance, and potentials, with the challenges, strategies, and models for easy application of IoT in different sectors and policy implications.

Keywords IoT · Privileged sector · Unprivileged sector

1 Introduction

In many industries, including the power sector, the automotive industry, safety and surveillance, remote health management, agriculture, smart homes, and smart cities, among others, IoT/M2M technology is being used to build smart infrastructure. In order to prevent harm to businesses, organizations, countries, and, most importantly, individuals, it is crucial to secure the Internet of things ecosystem end-to-end or from

A. Singh (✉) · V. Santhosh · K. W. Bavachan
Department of Commerce, Kristu Jayanti College (Autonomous), Bengaluru, India
e-mail: arti@kristujayanti.com; santhosh@kristujayanti.com; kenneth@kristujayanti.com

J. M. K. P. et al. (eds.), *Business Resilience and Digital Technology in the Post-Pandemic Era*, Contributions to Environmental Sciences & Innovative Business Technology, https://doi.org/10.1007/978-3-031-48075-1_21

the devices to the apps. Services could stop working as a result of a corrupted IoT network, causing confusion and panic. In order to succeed in this market, linked IoT devices must have end-to-end security; otherwise, IoT would cease to exist. In addition to security, the protection of personal information is another crucial concern, particularly in industries such.

The global Internet of things (IoT) security market is anticipated to increase from USD 8.2 billion in 2018 to USD 35.2 billion by 2023, at a CAGR of 33.7% over the course of the forecast year, according to a market research analysis released by MarketsandMarkets.

1.1 IoT and Its History

A network of actual physical objects that are linked to the Internet is known as the "Internet of things." These things are able to gather and exchange information about themselves and their environment. The Internet of things is already having a significant impact on a number of sectors, including manufacturing, transportation, healthcare, and retail.

Kevin Ashton is a co-founder of the MIT Auto-ID Center. He was the first to use the Internet of things in a 1999 presentation he presented to Procter & Gamble (P&G). The Internet of management was the topic of his lecture. Neil Gershenfeld, an MIT professor, in his 1999 book, *When Things Start to Think*, had written a reference to this. He didn't use the exact phrase but expounded on the direction IoT was taking. The Internet of things (IoT) has developed over the years, and there are now a variety of IoT products on the market, including smart homes, smartphones, watches, fire alarms, fitness trackers, and medical sensors.

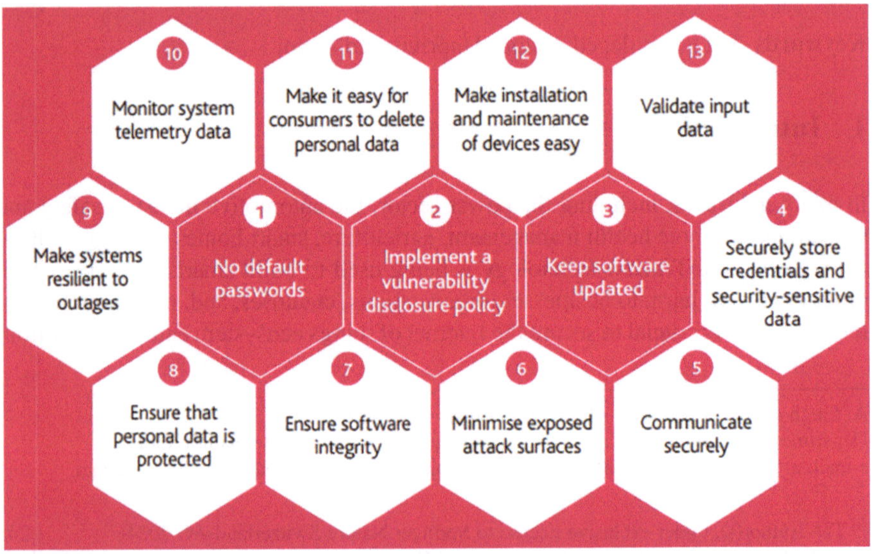

Source: https://assets.publishing.service.gov.uk/government/uploads/system/uploads

2 The Objectives of the Study

This chapter's objective goal is to identify and study in-depth the best research practices for IoT application methods and approaches. It examines the development and growth of IoT as well as its use in a number of application areas. The primary IoT application domains examined in this report are the environmental, smart city, commercial, industrial, and infrastructure domains. The study's secondary goals for providing IoT in India's rich and impoverished sectors are:

- To evaluate the current state of IoT adoption in both sectors
- Identify the benefits and challenges of implementing IoT
- Recommend strategies for expanding and enhancing IoT deployment to bridge the digital divide and promote inclusive growth in India.

3 Scope of the Study

The scope of Internet of things (IoT) technology in India is vast and has the potential to impact a wide range of sectors and industries. However, the impact and benefits of IoT are not equally distributed between privileged and unprivileged sectors in India. In privileged sectors such as urban areas and high-income communities, the scope of IoT is extensive and includes areas such as smart homes, connected vehicles, and industry 4.0. These communities have the infrastructure and resources to take full advantage of IoT technology, enabling them to improve efficiency, reduce waste, and enhance the quality of life. In unprivileged sectors such as rural areas and low-income communities, the scope of IoT is more limited. However, there is still significant potential for IoT to improve the quality of life in these communities. For example, IoT can be used to provide essential services such as healthcare and education, as well as improve agricultural productivity and reduce food waste. The scope of IoT in India is vast and has the potential to impact a wide range of sectors and industries. However, there are disparities in the extent to which different communities are able to take advantage of this technology, with privileged sectors having a wider scope of IoT compared to unprivileged sectors.

- *Adoption rate*: This column could compare the rate of IoT adoption between privileged and unprivileged sectors in India. For example, it could show that urban areas have a higher adoption rate compared to rural areas.
- *Infrastructure*: This column could compare the infrastructure available for IoT in privileged and unprivileged sectors in India. For example, it could show that urban areas have better infrastructure for IoT, including more widespread Internet connectivity and access to technology.
- *Digital literacy*: This column could compare the level of digital literacy between privileged and unprivileged sectors in India. For example, it could show that

urban areas have higher levels of digital literacy, enabling them to take advantage of IoT-based solutions.

- *Affordability:* This column could compare the affordability of IoT technology and services between privileged and unprivileged sectors in India. For example, it could show that IoT-based services are more expensive in rural areas, making them less accessible to low-income communities.
- *Government initiatives*: This column could highlight initiatives by the government to promote IoT adoption in unprivileged sectors in India. For example, it could describe government programs to provide Internet connectivity to rural areas or initiatives to increase digital literacy in these communities.
- *Private initiatives*: This column could highlight initiatives by private companies to provide IoT technology and services in unprivileged sectors in India. For example, it could describe projects to develop low-cost IoT devices for use in agriculture or initiatives to provide IoT-based healthcare services in rural areas.

4 Literature Survey

A number of authors have studied the provision of Internet of things (IoT) technology in India, focusing on the disparities between privileged and unprivileged sectors.

One author, Jain, highlights the significant gap in IoT adoption between urban and rural areas in India, with urban areas having much higher levels of access to and usage of IoT technology. Jain also notes that this gap is perpetuated by a lack of infrastructure and digital literacy in rural areas, which limits the ability of rural communities to take advantage of IoT-based solutions.

A group author, Chaturvedi et al., emphasizes the potential of IoT to bridge the digital divide and improve the quality of life in rural areas. The authors suggest that IoT can be leveraged to provide essential services such as healthcare and education in rural areas, as well as improve agricultural productivity and reduce food waste.

A third group of authors, Bhatnagar et al., also highlights the importance of IoT in improving the quality of life in rural areas. The authors suggest that IoT can be used to monitor and control water usage, power consumption, and other resources, which can help reduce waste and improve efficiency.

In conclusion, these authors all highlight the significant disparities in IoT provision between privileged and unprivileged sectors in India and the potential for IoT to help bridge the digital divide and improve the quality of life for all communities. However, the authors also acknowledge that there are significant barriers to overcome, including lack of infrastructure and digital literacy, to ensure that IoT is accessible and beneficial to all communities in India.

5 Data Protection Provisions for Consumer IoT

Personal data is processed by a lot of consumer IoT devices. Manufacturers should incorporate security elements that enable the protection of such personal data into consumer IoT devices. Additionally, there are rules and guidelines that deal with the protection of personal data in IoT devices used by consumers. For instance, the Personal Data Protection Bill 2018 of India, or any other applicable data protection law, must be followed by any devices and services processing personal data in India. The goal of the current document is to assist makers of consumer IoT devices in adding measures for the technological security of personal data.

1. For each device and service, the manufacturer must inform customers in a clear and transparent manner what personal data is processed, how it is used, by whom, and for what reasons. This also holds true for potential engaged third parties, such as marketers.
2. Consumer authorization must be sought legally whenever personal data is processed with their permission. Giving customers a free, obvious, and explicit opt-in choice as to whether their personal data can be used for a certain purpose is typically required to obtain consent "in a valid fashion."
3. Consumers who gave their consent to the processing of their personal data will be able to do so at any time. Customers anticipate being able to protect their privacy by properly configuring IoT device and service functioning.
4. The processing of personal data from consumer IoT devices and services should be limited to what is required to achieve the functional goals.
5. Consumers must be informed on what telemetry data is collected, how it is utilized, by whom, and for what purposes if it is gathered from consumer IoT devices and services.

6 Provision of IoT

Internet of things (IoT) availability in India differs across rich and underprivileged sectors. IoT technology is widely available and used for a range of purposes, including smart homes, transportation, and industrial automation, in privileged sectors, such as urban regions and large enterprises. Contrarily, underprivileged industries like small enterprises and rural areas have limited access to IoT technology and may encounter difficulties implementing it due to a lack of technical know-how, infrastructure, and resources. However, the Indian government has launched a number of programs to encourage IoT use in underserved sectors in an effort to close the digital divide.

India is using the Internet of things (IoT) in a number of industries, in both rich and underprivileged areas. IoT is being used in privileged areas to increase productivity and automate processes in sectors including manufacturing, transportation, and healthcare. For instance, IoT-enabled devices and sensors are being used to track and optimize logistics, remotely monitor patients' vital signs, and monitor and control industrial processes.

IoT is being utilized to increase access to vital services like healthcare, education, and energy in underdeveloped areas. For instance, IoT-enabled gadgets and sensors are being utilized to deliver remote medical assistance, expand educational accessibility through online courses, and track and manage energy use in off-grid regions.

6.1 An Architecture in IoT

The components of a hybrid IoT architecture are the things involved, the insight data processes, and the action that has to be taken (as shown in the figure below). It has the advantage of being able to include various subsystem architectures. The following is a list of each component's functions in a typical IoT design, as depicted in this figure:

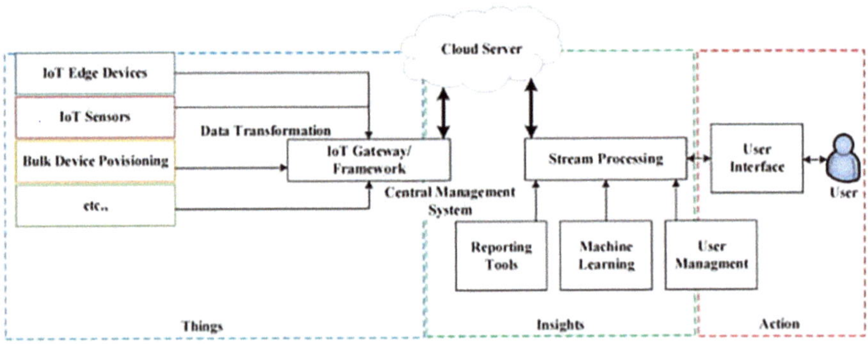

- Due to their capacity for self-conducted processing, IoT edge devices serve as the smart IoT actuator.
- IoT cloud, where data can be transmitted and received, is connected to the sensors.
- Connecting numerous devices to be registered is made easier with the aid of device provision.
- IoT gateway/framework acts as a cloud hub for IoT devices and offers command, management, and control over the devices.
- Time windowing reductions, stream aggregation, and external source combing are methods used in stream processing to examine complex execution.
- Extreme data can be used to predict and run the algorithms, thanks to machine learning. According to different conditions, it also analyses and facilitates preventive maintenance.
- While providing the essential tools for batch processing, reporting tools aid in holding and storing data and user management controls, which users or groups

are allowed to conduct an operation on the device. The procedure makes use of each user's application's capabilities.

6.2 Sector-Wise Application of IoT

Reporting IoT applications presented in recent studies is the main topic of this section of the study. A comprehensive taxonomy of IoT in the key application categories is shown in above figure. Healthcare, the environment, smart cities, business, industry, and infrastructure disciplines are the main application areas. What propels and explains the growth of this new trend, which ultimately results in the adoption of IoT by the modern world, are the applications and uses of IoT in various sectors. Studying IoT applications advances our knowledge of and ability to use IoT technology, which in turn facilitates the design of new systems for novel use cases. IoT can be defined as the generation of daily information from one object and its transmission to another. As a result, facilitating communication between things broadens, diversifies, and expands the scope of IoT applications.

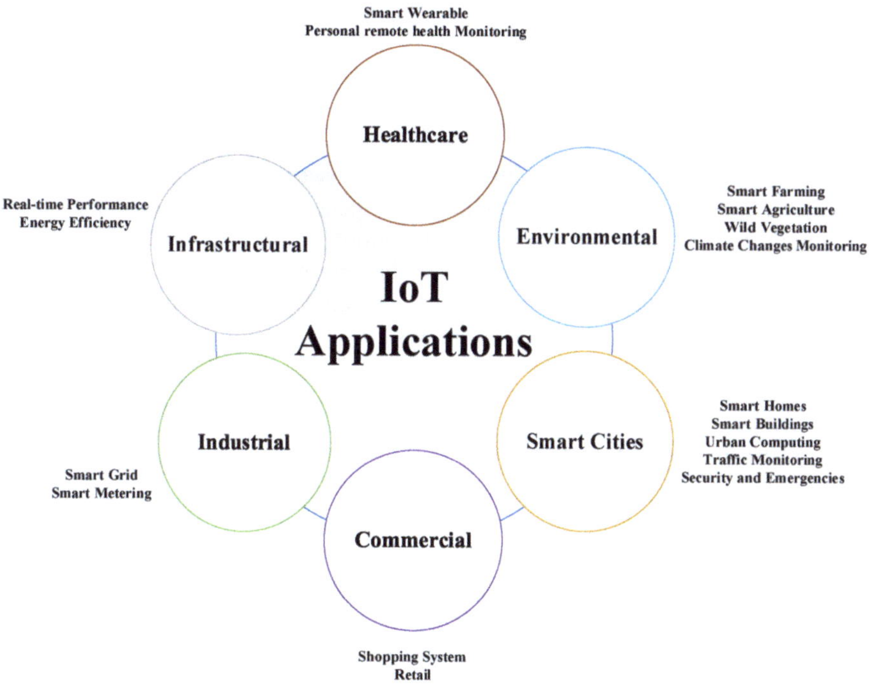

6.3 Healthcare Applications

According to the study, there are some important characteristics that have a bigger impact on users and help them verify healthcare services. The process involves carrying out analysis to assess the specified factors, such as risk sensitivity and trust, in order to qualify and validate the services.

6.4 Environment Applications

To assess the usage of IoT in an environmental context, numerous studies have been carried out. Li et al. first created an online monitoring system using a WSN for henhouses to monitor the environmental factors, such as the humidity, temperature, carbon dioxide, and ammonia (NH3). According to the authors, the bulk of applications in earlier studies had to do with the creation of systems, and the dependability of data transmission had not been taken into account.

6.5 Smart City Applications

IoT applications for smart city utilization have grown in number. A SenSquare structural design was put up by Montori et al. to observe consumer tests of existing data flows for the smart city using a mobile sensing crowd. The proposed facility structure made use of the data mining categorization procedure to assess the viability of the suggested strategy. Scalability cannot be approved because user privacy and data quality were not assessed. For a specific numerical forensics experimental prototype, Zia et al. presented a software design.

6.6 Commercial Applications

Commercial IoT applications have grown significantly in recent years. Alodib developed a prototype-driven method for automating the QoS-aware facility structure that involves real-time control. The author expressed concern over service-level agreement (SLA) violations. As a result, by specifying users, the SLAs were plotted to link the Petri Net model with the UML QoS model. It is crucial to analyze the discrete event system (DES), and this study revealed that the proposed strategy might give a combination facility that could be used in the Petri Net model. In order to fulfil QoS, it produced a review for assessing the implementation's cost-effectiveness; scalability, however, was not taken into account in the research.

6.7 Industrial Applications

In the area of industrial business, IoT technology has a number of significant applications. The usage of IoT in this field has been the subject of a number of important research. In order to improve the QoS through top-down decision-making processes while applying the Markov decision approach, Li et al. presented a three-stage planning methodology. Using MATLAB software, the authors created the scenario model and evaluated the suggested course of action. During the assessment process, a number of QoS parameters were considered, including the response time, end-to-end delay, accessibility, and bandwidth.

6.8 Infrastructural Applications

IoT-based networks can be employed simultaneously in a number of applications. As a result, this section highlights the research that have been done about the use of IoT in infrastructure. For IoT/Fog and SDN, Diro et al. developed an integrated structure. Significant benefits like reduced packet delay, the chance of missing packets, excess space conflict, and the highest throughput are made possible thanks to this. The divergence and necessity of flow space allocation, which can result in both necessary and dangerous flows, are stressed in this study. It illustrates how the Fog communication paradigm balances programmability of SDN processes with regular packet flows.

6.9 Relevance of IoT in Various Sectors

The potential of IoT to gather information about numerous facets of our lives is one of its most significant features. Numerous ways can be found to make our lives better with this data. Healthcare professionals can utilize IoT data, for instance, to track patient health, spot potential health problems, and deliver more individualized care. IoT data can be used by transportation companies to plan more effective routes, avoid traffic, and anticipate maintenance requirements. Additionally, IoT data can be used by manufacturers to streamline operations and cut waste.

6.10 Potentials of IoT

The Internet of things has enormous potential, especially in underdeveloped areas where data can be leveraged to address some of the most urgent issues. For instance, IoT can be utilized in the healthcare industry to increase patient outcomes, decrease inequities in care, and enhance access to care. IoT can be used in the transportation industry to increase security, lower emissions, and boost system productivity. IoT can also be used in the manufacturing industry to enhance worker safety, cut down on waste, and improve quality control.

7 Impacts of IoT in Various Underprivileged Sectors

IoT is already having a big impact on various industries, and its potential is particularly relevant in underprivileged industries. Data collected by IoT devices can be used to address some of the most pressing challenges in these sectors.

1. The Internet of things is an ecosystem of interconnected physical objects and devices that enables a network of sensors that have the ability to collect and exchange data.

This ecosystem of connected devices has the potential to revolutionize the way we live and work by changing the way we interact with the physical world around us.

In India, IoT is already being used in a number of different sectors, including agriculture, transportation, healthcare, and manufacturing. However, there is still a long way to go to realize the full potential of IoT in India.

One of the biggest challenges facing IoT adoption in India is the lack of standardization. With so many different devices and platforms available, it can be difficult to know which one to choose. This problem is compounded by the fact that many devices used in India are not compatible with each other.

Another problem is the lack of qualified personnel. India will need to upskill its workforce to take full advantage of IoT. This will require significant investment in training and education as well as developing new skills.

Despite these challenges, the potential of IoT in India is huge. By harnessing the power of the Internet of things, India has the opportunity to transform itself into a leading economic powerhouse.

2. The potential of IOT in India is huge, given the country's large population and vast geographical area.

The potential of IOT in India is indeed huge. With over 1.3 billion people and an area of over three million square kilometers, this country has the population density and geographic area for IOT to be a real game changer.

A number of pilot projects are already underway in India to test the potential of IOT. In one such project, farmers in the state of Maharashtra are using IOT-enabled sensors to monitor soil moisture levels, weather, and crop health. This data is then transmitted to a central server, where it can be analyzed to help farmers make more informed decisions about irrigation, planting, and crop care.

Other potential applications of IOT in India include reducing congestion and pollution in cities, improving supply chain management, and increasing the efficiency of agricultural production. With the right infrastructure and support, IOT has the potential to change many aspects of life in India for the better.

3. IOT has the potential to transform the underprivileged sector in India by increasing productivity and efficiency.

The underprivileged sector in India is a large part of the population that is not privy to the same resources and opportunities as the privileged sector. IOT has the potential to bring transformational change in the underprivileged sector by increasing productivity and efficiency.

IOT can help increase productivity and efficiency in the underprivileged sector in many ways. For example, IOT can be used to monitor and track resources and inventory, as well as automate tasks. This can free up time and resources that can be used more productively elsewhere. In addition, IOT can be used to improve communication and coordination between various stakeholders. This can lead to more efficient use of resources and a better overall outcome for the underprivileged sector.

The implementation of IOT in the underprivileged sector has the potential to bring about a transformational change that can improve the lives of those affected. IOT can help increase productivity and efficiency and improve communication and coordination. These changes can have a positive impact on the underprivileged sector and help improve the quality of life of those who live there.

4. IOT can help in efficient use of resources and waste management.

IoT can help in efficient use of resources and waste management. In particular, it can help with identifying and tracking sources, planning waste collection, and monitoring environmental conditions. By reducing the need for manual entry and increasing the accuracy of information, IoT can help optimize the use of resources and reduce waste.

5. IOT can also help with the safety and security of people in the underprivileged sector.

The Internet of things can help improve the safety and security of people in a number of underprivileged industries.

The Internet of things can help improve the safety and security of people in unprivileged sectors in a number of ways. For example, it can help them keep track of their belongings and property, and it can also help them stay informed about potential dangers in their environment. Additionally, the use of IoT devices can help unprivileged sector workers stay connected with each other and with emergency services in case of an accident or incident.

6. The implementation of IOT in the unprivileged sector.

The unprivileged sector in India has not been left out in the race to adopt and implement Internet of things (IoT). The government is taking various steps to ensure that the unprivileged sector is not left behind in the digital revolution. Various schemes and programs are being launched to promote the use of IoT in this sector. For example, the government has launched the "Digital India" scheme, which aims to provide access to digital technology to all citizens of India. Under this scheme, various initiatives are being taken to promote the use of IoT in the unprivileged sector. One such initiative is the "IoT for All" program, which is aimed at providing access to IoT devices and services to all citizens of India. The government is also working on a scheme to provide free Wi-Fi hotspots in all public places. This will help in promoting the use of IoT devices and services among the unprivileged sector.

7.1 Enactment of IoT in Privileged and Unprivileged Sectors

One of the most important aspects of IoT is its ability to collect data about various aspects of our lives. This data can be used to improve our lives in a number of ways. For example, healthcare providers can use IoT data to track patients' health, identify potential health risks, and provide more personalized care. Transportation companies can use IoT data to plan more efficient routes, avoid traffic congestion, and predict maintenance needs. And manufacturers can use IoT data to improve production processes and reduce waste.

The potential of IoT is particularly relevant in unprivileged sectors, where data can be used to address some of the most pressing challenges. For example, in the healthcare sector, IoT can be used to improve access to care, reduce disparities in care, and improve outcomes for patients. In the transportation sector, IoT can be used to improve safety, reduce emissions, and improve the efficiency of the transportation system. And in the manufacturing sector, IoT can be used to improve quality control, reduce waste, and improve worker safety.

7.2 Challenges of IoT

IoT is already having a major impact on various sectors, and its potential is particularly relevant in unprivileged sectors. Data collected by IoT devices can be used to address some of the most pressing challenges in these sectors.

1. The Internet of things is an ecosystem of interconnected physical objects and devices, enabled by the network of sensors, which have the ability to collect and exchange data.
2. This ecosystem of connected devices has the potential to revolutionize the way we live and work, by transforming the way we interact with the physical world around us.
3. In India, the IoT is already being used in a number of different sectors, including agriculture, transportation, healthcare, and manufacturing. However, there is still a long way to go before the full potential of the IoT can be realized in India.
4. One of the biggest challenges facing the adoption of the IoT in India is the lack of standardization. With so many different devices and platforms available, it can be

difficult to know which one to choose. This problem is compounded by the fact that many of the devices that are being used in India are not compatible with each other.

5. Another challenge is the lack of skilled personnel. In order to fully reap the benefits of the IoT, India will need to upskill its workforce. This will require a significant investment in training and education, as well as the development of new skillsets.

7.3 *Strategies for IoT in India: Privileged Vs Unprivileged Sectors*

There are several strategies that can be implemented to promote the adoption of the Internet of things (IoT) in India, particularly for privileged and unprivileged sectors.

For privileged sectors, such as urban areas, businesses, and government organizations, the focus can be on implementing IoT solutions to improve efficiency, reduce costs, and increase revenue. This can include the use of sensors, smart devices, and analytics to optimize operations, streamline processes, and gain insights into customer behavior.

For unprivileged sectors, such as rural areas and low-income communities, the focus can be on using IoT to address basic needs and improve quality of life. This can include the use of IoT-enabled devices to monitor and manage resources, such as water and energy, and to improve healthcare and education services.

Another approach could be a combination of both, where IoT enabled solutions are deployed in the privileged sectors and the learnings from the same are applied to unprivileged sectors to enable them to have the same or similar benefits.

In both cases, it is important for the government to provide support for IoT development and adoption through policies, regulations, and investment in infrastructure and education. Additionally, partnerships between government, private industry, and academic institutions can be formed to drive innovation, research, and development in IoT.

There are several strategies that can be implemented for IoT in India, depending on the sector in question.

For privileged sectors such as urban areas, the government can focus on building robust infrastructure and providing incentives for companies to develop and implement IoT solutions. This can include investments in 5G networks, data centers, and cloud computing services to support IoT applications. Additionally, the government can work with private companies to develop and implement smart city solutions, such as smart transportation systems and smart buildings.

For unprivileged sectors such as rural areas, the government can focus on implementing IoT solutions that address specific needs, such as improving agricultural productivity and increasing access to healthcare. This can include the use of IoT-enabled devices to monitor crop health and weather conditions, as well as telemedicine solutions that connect patients in remote areas with healthcare providers. Additionally, the government can provide training and education programs to help farmers and rural communities to adopt and benefit from IoT technology.

Overall, the key to successful IoT implementation in India will be a combination of both top-down and bottom-up approaches, with the government working in partnership with private companies and communities to develop and implement solutions that meet the unique needs of each sector.

8 Models for Easy Application of IoT

There are several models that can be used to make the application of IoT technology more accessible and easy for both privileged and unprivileged sectors in India. Some examples include the following:

Cloud-Based IoT Platforms These platforms can be used to connect, manage, and analyze data from a large number of IoT devices. This can help reduce the cost and complexity of deploying and maintaining IoT systems and make it more accessible for small- and medium-sized enterprises.

Low-Cost IoT Devices By developing low-cost IoT devices that are easy to use and maintain, it can help make IoT more accessible to a wider range of users, including those in the unprivileged sector.

Open-Source Solutions The use of open-source solutions for IoT can help reduce the cost and complexity of deploying and maintaining IoT systems and make it more accessible for small- and medium-sized enterprises.

Public-Private Partnership The government can partner with private companies to develop and deploy IoT solutions in the unprivileged sector. This can help leverage the expertise, resources, and funding of the private sector to address the challenges faced by the unprivileged sector.

Community-Based Approach Community-based approach can be used to deploy IoT solutions in unprivileged sectors, where the community members are trained to maintain and operate the devices, and the data is owned and controlled by the community.

Incubation Centers Incubation centers can be set up to provide support to start-ups and small businesses working in the IoT space. These centers can provide funding, mentoring, and other forms of support to help these companies develop and bring their products to market.

It's important to note that each of these models has its own advantages and disadvantages, and the best approach will depend on the specific needs and conditions of the target population. It's also important to evaluate the feasibility and sustainability of the model before implementing it.

Currently, there isn't a specific model of IoT that is widely used in India for both privileged and unprivileged sectors. However, there are several models that have been implemented or proposed in different parts of the country. The best model will depend on the specific needs and conditions of the target population.

One popular model that has been used in India is the public-private partnership model, where the government partners with private companies to develop and deploy IoT solutions in the unprivileged sector. This can help leverage the expertise, resources, and funding of the private sector to address the challenges faced by the unprivileged sector.

Another model that has been proposed is the community-based approach, where IoT solutions are deployed in unprivileged sectors, and the community members are trained to maintain and operate the devices, and the data is owned and controlled by the community. This approach can help ensure the sustainability of the IoT solutions and the community members have more ownership and control over the technology.

Models can be represented by the following

Public-private partnership model:

Government (or Public sector) <----- Partnership ----> Private sector

Community-based approach:

Community (or unprivileged sector) <----- IoT solution -------> Government/ Private sector (or privileged sector)

Both models have their own advantages and disadvantages, and it's important to evaluate the feasibility and sustainability of the model before implementing it. It's also important to take into account the specific needs and conditions of the target population and to involve all the stakeholders in the implementation process.

8.1 Policy Implications for IoT in India for Privileged and Unprivileged Sector

There are several policy implementations that can be put in place to support the growth and development of the IoT ecosystem in India for both privileged and unprivileged sectors. Some examples include:

- *Investing in infrastructure*: The Indian government can invest in building the necessary infrastructure to support IoT deployment, such as building out networks and data centers, and providing subsidies for IoT-enabled devices.
- *Encouraging innovation:* The government can create an environment that encourages innovation and entrepreneurship in the IoT space by providing funding, tax incentives, and other forms of support for start-ups and small businesses working in the IoT space.
- *Developing standards and regulations:* To ensure the security and privacy of IoT-enabled devices, the government can develop standards and regulations that govern the use of IoT technology.

- *Building capacity:* The government can invest in building the technical capacity of the country by providing training and education programs for IoT engineers and researchers.
- *Creating awareness:* The government can create awareness about the benefits of IoT for the unprivileged sector and encourage the use of IoT-enabled devices and services to improve access to education, healthcare, and other basic services in remote and underserved areas.
- *Promoting public-private partnership:* The government can encourage public-private partnerships to develop and deploy IoT solutions, with a focus on the unprivileged sector.

However, it's important to note that these policy implementations should be done with a long-term vision, taking into account the current and future needs of the society. It's also crucial to have a well-defined strategy, which can be executed seamlessly by taking all the stakeholders on board.

Specific IoT Applications in Privileged and Underprivileged Sectors In India

Privileged sectors	Underprivileged sectors
Smart homes and buildings	Precision agriculture and farming automation
Industrial automation and predictive	Water quality monitoring and management
Maintenance	Livestock monitoring and management
Smart retail and supply chain management	Community sanitation and waste management
Healthcare monitoring and telemedicine	Disaster management and emergency response

These IoT applications are designed to address specific challenges and opportunities in different sectors, ranging from increasing efficiency and productivity to promoting sustainability and inclusivity. By leveraging the power of IoT, India can bridge the digital divide and promote equitable growth across all sectors of the economy.

8.2 *Future of IoT in India for Privileged and Unprivileged Sectors*

The Internet of things (IoT) has the potential to bring significant benefits to both privileged and unprivileged sectors in India. In the privileged sector, IoT can help improve efficiency and productivity in industries such as manufacturing, transportation, and healthcare. For example, IoT-enabled devices can be used to track and optimize supply chains, monitor and maintain industrial equipment, and remotely monitor patients' health.

In the unprivileged sector, IoT can help address some of the country's most pressing social and economic challenges, such as improving access to education, healthcare, and basic services. For example, IoT-enabled devices can be used to deliver educational content to remote and underserved areas, monitor and improve the quality of water and air, and provide remote healthcare services to people in rural areas.

However, the implementation of IoT in India is still in the early stages, and there are several challenges that need to be addressed, such as lack of infrastructure and technical expertise, data privacy and security concerns, and a lack of standardization. To fully realize the potential of IoT in India, it will be important to address these challenges and to develop policies and regulations that support the growth and development of the IoT ecosystem.

9 Conclusions

In conclusion, the enactment of IoT in privileged and underprivileged sectors in India has the potential to bring big changes. However, it is important to consider the many factors that come into play for this to happen. There is no one-size-fits-all solution, and the right approach will vary by industry. However, with careful planning and execution, IoT can transform India for the better.

This study aimed to explain and outline emerging developments in IoT applications. The most recent research on IoT applications in the most crucial areas, such as healthcare, the environment, smart cities, and commercial and industrial application domains, is surveyed in this chapter. In order to comprehend the varied techniques, this study set out to review and assess the most popular IoT application domains. Data privacy and scalability for healthcare applications, cost and authorization issues for environmental applications, mobility and architectural issues for smart city applications, cost and implementation issues for commercial applications, hardware and production issues for industrial applications, and standardization and trust issues for infrastructure applications are just a few of the challenges that have been summarized in the study.

It has been said that in order to keep transaction information, improve the functionality of current structures, or create next-generation systems, a variety of IoT applications, including blockchain technology, still need to be utilized. Additional security, automatic business administration, dispersed platforms, offline-to-online information authentication, and other benefits may be attained as a result. Furthermore, the IoT's security and privacy features are the primary elements that may enable it to eventually become a technology that is widely used. However, IoT is vulnerable to multiple internal and external adversaries due to its self-organizing and accessible nature. This could jeopardize user security and privacy, opening the door for eavesdropping, access to personal information, and financial harm.

Consequently, to ensure data privacy, more sophisticated optimized algorithms and protocols are needed. The next generation of development technology can be created by developing an intelligent network that is both economical and energy-efficient and has the ability to support IoT system company growth. In India, there is a mismatch in the availability of IoT technology between rich and underprivileged sectors. Efforts are being made to close this gap and give all communities access to the advantages of IoT. The Internet of things, sometimes known as IoT, is a network of actual physical things that are linked to the Internet. These things are able to gather and exchange information about themselves and their environment. Healthcare, transportation, manufacturing, and retail are just a few of the industries that are already being significantly impacted by IoT.

References

1. G. S. Saini, S. Singh, N. Kumar (2018). Internet of Things (IoT) and smart cities: A survey of Indian scenario.
2. A. J. Jeyakumar and V. Srivastava (2020). Internet of things (IoT) for rural development in India: Opportunities, challenges and future directions.
3. V. Tripathi, A. Singh (2021). Internet of things (IoT) and its impact on the Indian economy.
4. S. Ghosh, S. Kumar, R. K. Jana (2018). IoT adoption in India: Opportunities and challenges.
5. R. K. Sharma, R. R. Sharma (2018). The Internet of things: Opportunities and challenges for business and industry in India.
6. A. Kumar, A. Singh, and P. Kumar (2020). A survey of Internet of things applications in India.
7. S. Chattopadhyay, A. Mukherjee (2019). IoT in agriculture: A study on its impact in rural India.
8. P. R. Lalitha, P. S. Sathidevi (2018). Smart villages in India: An IoT-based approach.
9. S. Subramanian, G. Das, R. Sundararajan (2019). Design of an IoT-based water quality monitoring system for rural India.
10. Chettri, L., & Bera, R. (2019). A comprehensive survey on Internet of things (IoT) toward 5G wireless systems. *IEEE Internet Things Journal, 7*, 16–32.
11. Tun, S. Y. Y., Madanian, S., & Mirza, F. (2020). Internet of things (IoT) applications for elderly care: A reflective review. *Aging Clinical and Experimental Research, 33*, 855–867.
12. Redhu, S., Maheshwari, M., Yeotikar, K., & Hegde, R. M. (2018). Joint data latency and packet loss optimization for relay-node selection in time-varying IoT networks. In *Proceedings of the 24th annual international conference on mobile computing and networking, New Delhi, India, 29 October* (pp. 711–713).
13. De Almeida, I. B. F., Mendes, L. L., Rodrigues, J. J., & da Cruz, M. A. (2019). 5G waveforms for IoT applications. *IEEE Communications Surveys & Tutorials, 21*, 2554–2567.
14. Kim, S., & Kim, S. (2018). User preference for an IoT healthcare application for lifestyle disease management. *Telecommunication Policy, 42*, 304–314.
15. Fafoutis, X., Clare, L., Grabham, N., Beeby, S., Stark, B., Piechocki, R., & Craddock, I. (2006). Energy neutral activity monitoring: Wearables powered by smart inductive charging surfaces. In *Proceedings of the 2016 13th annual IEEE international conference on sensing, communication, and networking (SECON), London, UK, 27–30* (pp. 1–9).
16. Jimenez, F., & Torres, R. (2015). Building an IoT-aware healthcare monitoring system. In *Proceedings of the 2015 34th international conference of the Chilean computer science society (SCCC), Santiago, Chile, 9–13* (pp. 1–4).
17. Ding, Y., Gang, S., & Hong, J. (2015). The design of home monitoring system by remote mobile medical. In *Proceedings of the 2015 7th international conference on information technology in medicine and education (ITME), Huangshan, China, 13–15* (pp. 278–281).
18. Atlam, H. F., & Wills, G. B. (2020). IoT security, privacy, safety and ethics. In *Digital twin technologies and smart cities* (pp. 123–149). Springer.
19. Temglit, N., Chibani, A., Djouani, K., & Nacer, M. A. (2017). A distributed agent-based approach for optimal QoS selection in web of object choreography. *IEEE Systems Journal, 12*, 1655–1666.

20. Cao, B., Liu, J., Wen, Y., Li, H., Xiao, Q., & Chen, J. (2019). QoS-aware service recommendation based on relational topic model and factorization machines for IoT mashup applications. *Journal of Parallel and Distributed Computing, 132*, 177–189.
21. Cuomo, S., Di Somma, V., & Sica, F. (2018). An application of the one-factor HullWhite model in an IoT financial scenario. *Sustainable Cities and Society, 38*, 18–20.
22. Hull, J., & White, A. (1994). Numerical procedures for implementing term structure models I: Single-factor models. *Journal of Derivatives, 2*, 7–16.
23. Pustišek, M., & Kos, A. (2018). Approaches to front-end IoT application development for the Ethereum blockchain. *Procedia Computer Science, 129*, 410–419.
24. Park, J. H. (2019). Advances in future Internet and the industrial Internet of Things. *Symmetry, 11*, 244.
25. Li, L., Li, S., & Zhao, S. (2014). QoS-aware scheduling of services-oriented internet of things. *IEEE Transactions on Industrial Informatics, 10*, 1497–1505.
26. Venticinque, S., & Amato, A. (2019). A methodology for deployment of IoT application in fog. *Journal of Ambient Intelligence and Humanized Computing, 10*, 1955–1976.
27. Jin, Y., Raza, U., Aijaz, A., Sooriyabandara, M., & Gormus, S. (2017). Content centric cross-layer scheduling for industrial IoT applications using 6TiSCH. *IEEE Access, 6*, 234–244.
28. Kiran, M. S., & Rajalakshmi, P. (2018). Performance analysis of CSMA/CA and PCA for time critical industrial IoT applications. *IEEE Transactions on Industrial Informatics, 14*, 2281–2293.
29. Ahmad, S., Alam, N., & Hasan, M. (2018). Robust TFET SRAM cell for ultra-low power IoT applications. *AEU-International Journal of Electronics and Communications, 89*, 70–76.
30. Kwon, D., Hodkiewicz, M. R., Fan, J., Shibutani, T., & Pecht, M. G. (2016). IoT-based prognostics and systems health management for industrial applications. *IEEE Access, 4*, 3659–3670.
31. Luvisotto, M., Tramarin, F., Vangelista, L., & Vitturi, S. (2018). On the use of LoRaWAN for indoor industrial IoT applications. *Wireless Communications and Mobile Computing, 2018*, 1–11.
32. Mazzei, D., Baldi, G., Fantoni, G., Montelisciani, G., Pitasi, A., Ricci, L., & Rizzello, L. (2020). A blockchain tokenizer for industrial IOT trustless applications. *Future Generation Computer Systems, 105*, 432–445.
33. Diro, A. A., Reda, H. T., & Chilamkurti, N. (2018). Differential flow space allocation scheme in SDN based fog computing for IoT applications. *Journal of Ambient Intelligence and Humanized Computing, 1–11.*
34. Naranjo, P. G. V., Baccarelli, E., & Scarpiniti, M. (2018). Design and energy-efficient resource management of virtualized networked fog architectures for the real-time support of IoT applications. *The Journal of Supercomputing, 74*, 2470–2507.
35. Chen, X., Shi, Q., Yang, L., & Xu, J. (2018). ThriftyEdge: Resource-efficient edge computing for intelligent IoT applications. *IEEE Network, 32*, 61–65.
36. Mangia, M., Pareschi, F., Varma, R., Rovatti, R., Kovačević, J., & Setti, G. (2018). Rakeness-based compressed sensing of multiple-graph signals for IoT applications. *IEEE Transactions on Circuits and Systems II: Express Briefs, 65*, 682–686.
37. Taghadosi, M., Albasha, L., Quadir, N. A., Rahama, Y. A., & Qaddoumi, N. (2017). High efficiency energy harvesters in 65nm CMOS process for autonomous IoT sensor applications. *IEEE Access, 6*, 2397–2409.
38. Alabady, S. A., Salleh, M. F. M., & Al-Turjman, F. (2018). LCPC error correction code for IoT applications. *Sustainable Cities and Society, 42*, 663–673.
39. Ouedraogo, C. A., Medjiah, S., Chassot, C., & Drira, K. (2018). Enhancing middleware-based IoT applications through run-time pluggable Qos management mechanisms. Application to a oneM2M compliant IoT middleware. *Procedia Computer Science, 130*, 619–627.

Work Group Inclusiveness Influenced by Working Virtually in IT Sector, Bangalore

C. H. Raja Kamal, M. Chandrakala, Ponny Thomas, and P. Kumarasamy

Abstract In human resource management study, understanding how online work environments affect disenfranchised workers is important. Inclusivity academics have concentrated on how dissimilarities affect managerial inclusion of workers but seldom on how this happens via technology-mediated contact. We propose a theoretic model of two common distance types using inclusiveness and virtual work research on identity and communication (targeted and diffuse). We then explore how social distance impacts inclusion. Our data reveals that the virtual workplace may lessen certain prejudice and discrimination while boosting others. The results suggest inclusiveness studies should be more context-sensitive.

Keywords Human resource · Information technology · Socioeconomic

1 Introduction

Is the inclusiveness of today's work groups understandable without virtual work research? Academics studying inclusivity have mostly ignored how digital collaboration techniques are used in businesses [1]. For instance, the circumstance where employees are no longer coworkers in a central office has not yet had a substantial impact on the study of work group inclusion.

The literature on inclusion has helped organizations increase their talent pools [2], foster welfare and employee retention [3], and use diversity for innovation and performance [4]. Sadly, no solution has yet been proposed to explain why employee physical involvement is falling.

Inclusion has drawn the interest of several virtuality studies.

This study offers a theoretical framework for understanding how the virtual workplace influences the inclusion of underprivileged workers in work groups. According to what we suggest, workplace inclusion varies [5]. Instead, the

C. H. Raja Kamal (✉) · M. Chandrakala · P. Thomas · P. Kumarasamy
Department of Commerce, Kristu Jayanti College (Autonomous), Bangalore, India

J. M. K. P. et al. (eds.), *Business Resilience and Digital Technology in the Post-Pandemic Era*, Contributions to Environmental Sciences & Innovative Business Technology, https://doi.org/10.1007/978-3-031-48075-1_22

workplace environment has an impact on how engaged employees feel [6]. Inclusion in the workplace and the idea of virtual employment are combined in our theoretical framework. This overlap of research areas suggests that inclusion challenges may be influenced by online contexts. Below, we analyse the research supporting our conceptual model.

2 Remote Work

For our study, virtual work is work done remotely utilizing know-how ([7], p. 259). Employees have the option of working remotely and on the go.

There are many definitions of virtual work, but geographical dispersion and reliance on technology are common (For instance, [8, 9]). Communication's influence is crucial [10]. Technology-mediated communication is distinct from face-to-face conversation, and the lack of nonverbal indications like smiles and headshakes makes communication more challenging [11, 12]. It could be challenging to comprehend messages or the emotions driving them without these movements [13].

2.1 Inclusiveness

We place a strong emphasis on each person's inclusion in the team. The experiences of the marginalized individual with regard to being perceived as different and perhaps being an outsider are of particular concern to us, not least because core members are more likely to interact with and share expertise with outsiders than with other work group members [14, 15].

The phrase "degree to which an employee thinks that he or she is a respected member of the work group by receiving treatment that meets his or her requirements for belongingness and uniqueness" is taken from Shore et al.'s [16] definition of work group inclusion.

Based on the aforementioned, we split inclusivity into group-oriented and individual-oriented dimensions (Refer to [16, 20]). However, we combine socioemotional and task-oriented qualities (cf. [21, 22]). A matrix of four inclusiveness dimensions is shown in Table 1. showing disadvantaged people's positive experiences. This paradigm makes the inclusiveness idea more applicable for studying how a particular contextual environment (virtual work) influences various characteristics of perceived inclusion.

Table 1 Representing the factors of group and individual orientation

Factor	Socioemotional	Task related
Group oriented	Satisfy need for group membership and acceptance. Reciprocity and love	Information and resources required for work from the group. Participating in key group decisions
Individual oriented	Ensured individuality, sincerity, and self-worth. Voice recognition (e.g. [16, 17])	Use of the individual's particular knowledge, views, and skills. The individual influencing decision-making and work group functioning (e.g., [18, 19])

2.2 Model Development

Social identity theory describes the inclusion of diverse workgroup members [14]. This idea says people classify depending on similarities and contrasts, they and others see themselves as insiders or outsiders [23]. As a result, insiders are seen favourably and outsiders unfavourably.

Identification and categorization are not isolated processes. Hewstone et al. [6] suggest that context cue salience affects identification processes. According to Montoya and Horton [24], identification and dissimilarity depend on the expertise of the work group. Thus, relationships inside the workplace display both differences and similarities in appearance and conduct once identification procedures include different people. Technology-mediated communication has less information and vaguer clues about personality [12]. Hence, dissimilarity and inclusivity may interact differently in virtual workplaces.

Identification and communication must be examined to explain this. Wilson et al. [25] claim that a poor relationship between communication and group identification creates perceived social distance in virtual environments. Consequently, poor group communication is linked to low group identification. Dissimilarity and virtual labour may decrease the communication-identification relationship, increasing social distance [26]. Yet communication and identification may take diverse paths to social distance. Hence, dissimilarity and the virtual workplace facilitate perceived social distance. We refer to them as distributed social distance and targeted social distance (towards particular work group members) (of a more neutral character between all work group members).

2.2.1 Social Distance That Is Both Focused and Dispersed

A measure of empathy and proximity between people is called social distance [27]. This idea contends that individuals who experience social isolation are hesitant to interact with others [28]. In the virtual work literature, perceived distance [25], psychological distance [29], and social distance are all subjective perceptions of how close a person feels to others socially.

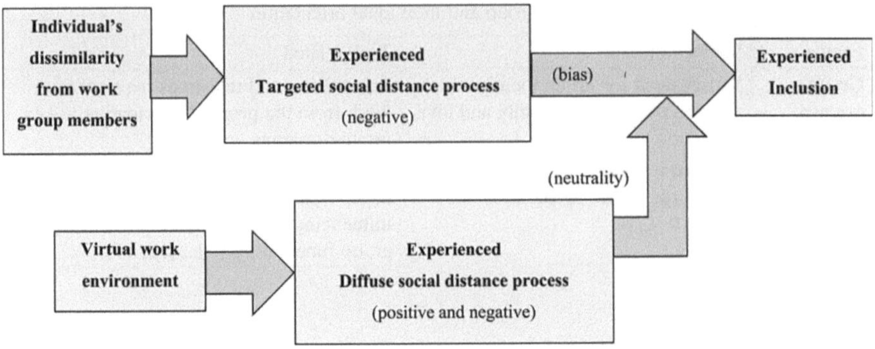

Fig. 1 A model of the interaction between targeted and diffuse social distance in relation

Perceived differences in characteristics make it difficult to identify with a group at work, which prevents communication with the group [30, 31]. Even though inclusion is crucial for all members of the work group, it has been said that it is a larger challenge when it comes to differences. According to Tawa et al. [32], social distance has been used to gauge hostility towards foreigners in an indirect manner. Being an outsider is the main route from dissimilarity to restricted communication. Hence, dominant group members don't interact with disadvantaged people since they don't identify with them. Hence, dissimilarity-based social distance targets underprivileged workers and negatively impacts inclusion (cf. [14]).

The virtual workplace primarily affects communication since it is mediated by technology. Announcement lacks social cues about the work group and is formal. This is due to how impersonal and unintentional Internet communication is [33, 34]. Due to poor interpersonal communication, the work group's identity is seriously impacted [35–37]. People are unable to appropriately appraise one another in the absence of signs. As proven by Armstrong and Cole [38]. Virtual work environments may diminish employee relationships, making it hard for distant workers to communicate. Vough [39] observed that online engagement decreased group familiarity and identification. Hence, minimal communication leads to poor identification from virtual labour. Hence, less interpersonal interaction is the major reason group members don't identify as strongly [29, 40]. Hence, whereas dissimilarity might contribute to focused social distance, a virtual work environment promotes diffuse social distance that is larger and more neutral.

This might hurt inclusivity but possibly improve it. In a virtual workplace, less social signals may mean work group members are regarded less prejudiced than predicted from a dissimilarity standpoint [41]. Stahl et al. [42] contend that social and demographic factors are less significant in online groups due to the leaner, more neutral communication. According to Velez-Calle et al. [43], technology-mediated communication may make discrimination less pronounced. Therefore, information technology lowers overt disparities and behavioural differences, removing some obstacles to contact between diverse people [44–47].

We provide a conceptual model (Fig. 1) that connects social identity dynamics and work group inclusion to informational cue salience based on the aforementioned literature study [12, 24]. The impact of targeted and scattered social distance on each inclusion parameter is our next suggestion.

According to Velez-Calle et al. [43], communication mediated by technology may lessen discrimination. As a result, information technology lessens overt disparities and behavioural variances, removing some barriers to contact between individuals with different backgrounds [44–46]. We provide a conceptual model (Fig. 1) that connects social identity dynamics and work group inclusion to informational cue salience based on the aforementioned literature study [12, 24]. The impact of targeted and scattered social distance on each inclusion parameter is our next suggestion.

2.2.2 Targeted-Diffuse Social Distance

This section discusses the effects of both concentrated and diffuse social distance on work group inclusion. This is carried out for each of the four sub-dimensions of inclusiveness: task-related group, task-related individual, socioemotional person, and socioemotional group. Our main presumption is that unfair treatment of disadvantaged persons will have a smaller positive and negative impact, thanks to the advent of the virtual workplace.

2.2.3 Socioemotional Group-Oriented Inclusion

Being a part of a group is crucial for feeling included, claim Shore et al. [16]. Since group belongingness sensations are essential to human life from an evolutionary perspective, people have a natural incentive to pursue and sustain them. Some claim that there are many areas where collective participation is required, with the workplace today playing a significant role [48]. The sense of community among employees is related to socioemotional group inclusion. According to Baumeister and Leary [49], in order to have a long-lasting, loving relationship with co-workers, one must have pleasant interactions with them.

Dissimilar people, who regularly experience unfavourable views and targeted social distancing, will find it harder to feel like part of the group since it depends on good, long-term interactions. Hence, different workers will have a weaker socioemotional bond to dominating workers.

Diffuse social distance, such as increasing neutrality and apathy, may not alleviate this problem in the virtual world. Social disengagement will rise due to physical distance and the necessity to communicate through technology [11, 12]. So fostering group cohesion in a virtual workplace may be tough. Diffuse social distance may decrease. All members of the work group should strive for socioemotional inclusion, but those who are already disadvantaged may suffer the most. Our first suggestion follows from the aforementioned.

Proposition 1 Targeted social distance negatively impacts socioemotional group-oriented inclusion, whereas diffuse social distance worsens it, inclusiveness with a socioemotional focus on the individual. An additional important socioemotional component of inclusion is feeling valued as an individual ([50]; [16]). A crucial component of human identity is being uniquely individual [51]. The private sense of self that people have is valued in both Western and Eastern cultures as being different from the community. Therefore, a desire for uniqueness and non-conformity, which some people think are necessary for enjoyment and a sense of purpose in life, is linked to the want to feel distinctive [52].

Pride in one's identity inside the workplace can boost self-confidence in marginalized and diverse individuals [53]. Pride in one's minority group improves self-esteem and resilience and lowers unhappiness [54, 55]. Yet purposeful social distancing may cause one to question their self-worth and their distinctive social function in the workplace.

Online, underprivileged workers might create positive self-images [45]. Diffuse social distance reduces the social need for conformity and status battles [56]. Minorities may benefit from online group neutrality. Furthermore, less close engagement may not be an issue. While intimate touch with others may both affirm and contradict one's self-perception, feeling favourably distinctive doesn't need it, hence our second proposal.

Proposition 2 Targeted social distance has a modestly unfavourable influence on socioemotional individual-oriented inclusion, whereas diffuse social distance will significantly offset it.

2.2.4 Task-Related Group-Oriented Inclusion

Members of the workgroup must have a connection to the task. Working with the group is a requirement for task-related inclusion. According to O'Hara et al. [57], feeling like a part of the group is dependent upon cooperation, information sharing, and decision-making. Pelled et al. [22] stressed that participation calls for access to the knowledge of the work group and group communication of the organization's aims.

Different workers have long struggled with access to information and collective decision-making because of prejudice and negative bias that disenfranchise them [19, 58]. The exclusion of underrepresented groups from important social networks [59] could be motivated by direct discrimination and other adverse biases [60]. Economic self-interest may prevent members of outgroups from advancing economically task-related tasks, which are often hindered by social distance. Our third suggestion follows.

Proposition 3 Targeted social distance negatively impacts task-related group-oriented inclusion, although diffuse social distance mitigates this.

2.2.5 Task-Based Inclusion

One way to become more inclusive is to use their distinctive ideas and skills in task-related activities. This intentional use of one's own resources is what gives people in disadvantageous situations self-confidence (cf. [61, 62]). Outsiders and minorities may provide unique viewpoints and information in addition to doing their job [18].

Yet digesting specialized knowledge in diverse groups may be a challenge. Due to prejudices against dissimilar persons, work group members can't always work well with others with diverse views and beliefs [5, 23]. So purposeful social distance may hinder problem-solving utilizing different people's distinct task-related traits.

The dispersed social distance in virtual work environments may also affect the sensation of being recognized for specialized job contribution. Communicating useful, complementary ideas may be harder when communication is constrained. However, increased general neutrality and dispersed social distance may make various group members feel supportive. Therefore, virtual workplaces may give disadvantaged workers more of a priority than ones in physical locations [44]. This might be the case because social neutrality in online work groups reduces conformity and groupthink, allowing people to express illogical opinions [56]. Internet neutrality may therefore enable group members to communicate unique and useful information [63]. Targeted social distance typically lowers task-related individual level inclusion for members of minority groups, although the virtual environment may increase this.

3 Discussion

HRM is discussing inclusivity. Workgroup communication and identification are key to inclusion [18, 19, 64] [16, 17, 65]. However, those two elements are impacted by working remotely [29, 40, 66]. This recommends a more in-depth discussion of inclusivity and virtual employment. In order to contribute to the literature, this chapter created a conceptual model related to inclusion.

Our model construction and stated propositions provide three reasons that need additional consideration. First, we need additional study on how social distance affects organizations and their interactions. Second, inclusion context requires further attention. This article emphasizes the expanding digitization of labour, although other organizational situations may also need more study. Finally, to examine contextual impact, we propose distinguishing between group and individual inclusion characteristics, socioemotional, and task-related variables.

4 Relevance to Research

A conceptual model of how targeted social distance works is the main contribution we make, caused by dissimilarity interacts with diffuse social distance in a virtual work environment to predict inclusion. We contend that dispersed social distance exacerbates the negative consequences of exclusion on socioemotional inclusion. We predict diffuse social distance to counteract the other inclusiveness characteristics, so dissimilar persons with outsider status will feel more included in the virtual workplace than the physical workplace. Future studies should consider how virtual inclusion affects different aspects. Our approach also suggests ways to study social distance outside of virtual employment. Inclusivity study has indicated that a strong professional hierarchy may discourage low-status workers from speaking up [5]. In bureaucratic organizations, strict rules and norms are another background [67]. Current research suggests studying how rules and norms affect inclusion practices [68]. It's pertinent to explore how strict laws and norms place communication and identification processes and how this may generate social distance between a person and their work group.

We've provided a paradigm for virtual job inclusion, but further study is needed. To provide the groundwork for future study, we outlined the existing structure. Our assertions might be quantitatively tested using our conceptual model. Qualitative research on targeted and diffuse social distance dynamics and feedback mechanisms is another option. A possible feedback between inclusion and the two social distance measures might be assessed here. Inclusion is likely to reduce social gap among coworkers. We have not incorporated this factor in the present model since we have studied the contextual function of the virtual world for outsider inclusion. Yet, this topic warrants more study.

Diffuse social distance may hinder minority inclusion more than others. Jackowska and Lauring [69] discovered that virtuality made people from communal cultures see their coworkers negatively. One may also hypothesize that collectivist cultures value being part of the community socioemotionally more than individualist cultures value being valued as an individual. Minorities may value other inclusivity factors. Linguistic minorities may have trouble accessing group networks and information, resulting in a task-related group connection [70]. Nishii [17] observed that gender-inclusive environments need active valuing of all viewpoints. Virtual organizations must investigate these inclusiveness issues. Inclusivity in hybrid workplaces where workers meet face-to-face more often is a last topic. Targeted and dispersed social distance may play distinct roles in hybrid workplaces than in virtual ones. HRM research on this topic is few.

5 Implications for Practice

Inclusion management advice for virtual workplaces is provided in this article. According to our concept, the absorption of different outsiders will have both positive and bad effects in the virtual world. Managers should stress the advantages of a decrease on the detrimental impacts of shaky group bonds and create a more formal, objective work atmosphere. Managers might promote regular and open communication [71, 72]. Others have asserted that policies and digital platforms might provide impartial access to information and decision-making, allowing the person to feel valuable and a part of the team from the beginning [73].

References

1. Asatiani, A., Hämäläinen, J., Penttinen, E., & Rossi, M. (2021). Constructing continuity across the organisational culture boundary in a highly virtual work environment. *Information Systems Journal, 31*, 62–93.
2. Festing, M., Kornau, A., & Schafer, L. (2015). Think talent–think male? A comparative case study analysis of gender inclusion in talent management practices in the German media industry. *International Journal of Human Resource Management, 26*, 707–732.
3. Mor Barak, M. E., Levin, A., Nissly, J. A., Lane, C., & J.). (2006). Why do they leave? Modeling child welfare workers' turnover intentions. *Children and Youth Services Review, 28*, 548–577.
4. Leroy, H. L., Buengeler, C., Veestraeten, M., Shemla, M., & Hoever, I. J. (2021). Fostering team creativity through team-derived inclusion: The role of leader harvesting the benefits of diversity and cultivating value-in-diversity beliefs. *Group and Organization Management, 47*, 798.
5. Nembhard, I. M., & Edmondson, A. C. (2006). Making it safe: The effects of leader inclusiveness and professional status on psychological safety and improvement efforts in health care teams. *Journal of Organizational Behavior: The International Journal of Industrial, Occupational and Organizational Psychology and Behavior, 27*, 941–966.
6. Hewstone, M., Rubin, M., & Willis, H. (2002). Intergroup bias. *Annual Review of Psychology, 53*, 575–604.
7. Raghuram, S., & Wiesenfeld, B. (2004). Work-nonwork conflict and job stress among virtual workers. *Human Resource Management Review, 43*, 259–277.
8. Gibson, C. B., & Gibbs, J. L. (2006). Unpacking the concept of virtuality: The effects of geographic dispersion, electronic dependence, dynamic structure, and national diversity on team innovation. *Administrative Science Quarterly, 51*, 451–495.
9. Kirkman, B. L., & Mathieu, J. E. (2005). The dimensions and antecedents of team virtuality. *Journal of Management, 31*, 700–718.
10. Marlow, S. L., Lacerenza, C. N., & Salas, E. (2017). Communication in virtual teams: A conceptual framework and research agenda. *Human Resource Management Review, 27*, 575–589.
11. Cramton, C. D. (2001). The mutual knowledge problem and its consequences for dispersed collaboration. *Organization Science, 12*, 346–371.
12. Hertel, G., Geister, S., & Konradt, U. (2005). Managing virtual teams: A review of current empirical research. *Human Resource Management Review, 15*, 69–95.
13. Ekman, P. (1993). Facial expression and emotion. *American Psychologist, 48*, 384.

14. Shore, L. M., & Chung, B. G. (2021). Inclusive leadership: How leaders sustain or discourage work group inclusion. *Group & Organization Management, 47*, 723.
15. Subasi, B., van Ginkel, W. P., & van Knippenberg, D. (2021). Minority status, access to information, and individual performance. *Journal of Applied Social Psychology, 51*, 159–175.
16. Shore, L. M., Randel, A. E., Chung, B. G., Dean, M. A., Holcombe Ehrhart, K., & Singh, G. (2011). Inclusion and diversity in work groups: A review and model for future research. *Journal of Management, 37*, 1262–1289.
17. Nishii, L. H. (2013). The benefits of climate for inclusion for gender-diverse groups. *Academy of Management Journal, 56*, 1754–1774.
18. Ferdman, B. M. The practice of inclusion in diverse organizations: Toward a systemic and inclusive framework, (2014).
19. Mor-Barak, M. E., Cherin, D. A., & Berkman, S. (1998). Organizational and personal dimensions in diversity climate. *The Journal of Applied Behavioral Science, 43*, 82–104.
20. Mor Barak, M. (2000). The inclusive workplace. *Social Work*, 1–12.
21. Jansen, W. S., Otten, S., van der Zee, K. I., & Jans, L. (2014). Inclusion: Conceptualization and measurement. *European Journal of Social Psychology, 44*, 370–385.
22. Pelled, L. H., Ledford, G. E., & Mohrman, S. A. (1999). Demographic dissimilarity and workplace inclusion. *Journal of Management Studies, 36*, 1013–1031.
23. Tajfel, H., & Turner, J. C. (1986). The social identity theory of intergroup behavior. In S. Worchel & W. G. Austin (Eds.), *Psychology of intergroup relations* (pp. 7–24).
24. Montoya, R. M., & Horton, R. S. (2013). A meta-analytic investigation of the processes underlying the similarity-attraction effect. *Journal of Social and Personal Relationships, 30*, 64–94.
25. Wilson, J. M., O'Leary, M. B., Metiu, A., & Jett, Q. R. (2008). Perceived proximity in virtual work: Explaining the paradox of far-but-close. *Organization Studies, 29*, 979–1002.
26. Storm, I., Sobolewska, M., & Ford, R. (2017). Is ethnic prejudice declining in Britain? Change in social distance attitudes among ethnic majority and minority Britons. *British Journal of Sociology, 68*, 410–443.
27. Magee, J. C., & Smith, P. K. (2013). The social distance theory of power. *Personality and Social Psychology Review, 17*, 158–186.
28. Foulk, T. A., De Pater, I. E., Schaerer, M., du Plessis, C., Lee, R., & Erez, A. (2020). It's lonely at the bottom (too): The effects of experienced powerlessness on social closeness and disengagement. *Personnel Psychology, 73*, 363–394.
29. Eisenberg, J., & DiTomaso, N. (2021). Structural decisions about configuration, assignments, and geographical distribution in teams: Influences on team communications and trust. *Human Resource Management Review, 31*, 100739.
30. Greenaway, K. H., Wright, R. G., Willingham, J., Reynolds, K. J., & Haslam, S. A. (2015). Shared identity is key to effective communication. *Personality and Social Psychology Bulletin, 41*, 171–182.
31. Mannix, E., & Neale, M. A. (2005). What differences make a difference? *Psychological Science in the Public Interest, 6*, 31–55.
32. Tawa, J., Negron, R., Suyemoto, K. L., & Carter, A. S. (2015). The effect of resource competition on Blacks' and Asians' social distance using a virtual world methodology. *Group Processes & Intergroup Relations, 18*, 761–777.
33. Gibbs, J. L., Rozaidi, N. A., & Eisenberg, J. (2013). Overcoming the "ideology of openness": Probing the affordances of social media for organizational knowledge sharing. *Journal of Computer-Mediated Communication, 19*, 102–120.
34. Johri, A. (2012). From a distance: Impression formation and impression accuracy among geographically distributed coworkers. *Computers in Human Behavior, 28*, 1997–2006.
35. Johnson, S. K., Bettenhausen, K., & Gibbons, E. (2009). Realities of working in virtual teams: Affective and attitudinal outcomes of using computer-mediated communication. *Small Group Research, 40*, 623–649.

36. Thatcher, S. M. B., & Zhu, X. (2006). Changing identities in a changing workplace: Identification, identity enactment, self-verification, and telecommuting. *Academy of Management Review, 31*, 1076–1088.
37. Wiesenfeld, B. M., Raghuram, S., & Garud, R. (2001). Organizational identification among virtual workers: The role of need for affiliation and perceived work-based social support. *Journal of Management, 27*, 213–229.
38. Armstrong, D. J., & Cole, P. (2002). Managing distances and differences in geographically distributed work groups. In P. Hinds & S. Kiesler (Eds.), *Distributed work* (pp. 167–186). MIT Press.
39. Vough, H. (2012). Not all identifications are created equal: Exploring employee accounts for workgroup, organizational, and professional identification. *Organization Science, 23*, 778–800.
40. Eisenberg, J., & Mattarelli, E. (2017). Building bridges in global virtual teams: The role of multicultural brokers in overcoming the negative effects of identity threats on knowledge sharing across subgroups. *Journal of International Management, 23*, 399–411.
41. Liberman, N., & Trope, Y. (2008). The psychology of transcending the here and the now. *Science, 322*, 1201–1205.
42. Stahl, G., Maznevski, M., Voigt, A., & Jonsen, K. (2010). Unraveling the effects of cultural diversity in teams: A meta-analysis of research on multicultural work groups. *Journal of International Business Studies, 41*, 690–709.
43. Velez-Calle, A., Mariam, M., Gonzalez-Perez, M. A., Jimenez, A., Eisenberg, J., & Santamaria-Alvarez, S. M. (2020). When technological savviness overcomes cultural differences: Millennials in global virtual teams. In *16. Critical perspectives on international business* (pp. 279–303).
44. Driskell, J. E., Radtke, P. H., & Salas, E. (2003). Virtual teams: Effects of technological mediation on team performance. *Group Dynamics: Theory, Research, and Practice, 7*, 297.
45. Han, S. J., & Beyerlein, M. (2016). Framing the effects of multinational cultural diversity on virtual team processes. *Small Group Research, 47*, 351–383.
46. Lauring, J., & Klitmøller, A. (2015). Corporate language-based communication avoidance in MNCs: A multi-sited ethnography approach. *Journal of World Business, 50*, 46–55.
47. Selmer, J., Dickmann, M., Froese, F. J., Lauring, J., Reiche, B. S., & Shaffer, M. (2021). The potential of virtual global mobility: Implications for practice and future research. *Journal of Global Mobility, 10*, 1–13.
48. Leary, M. R., & Cox, C. B. (2008). Belongingness motivation: A mainspring of social action. In J. Y. Shah & W. L. Gardner (Eds.), *Handbook of motivation science* (pp. 27–40). Guilford Press.
49. Baumeister, R. F., & Leary, M. R. (1995). The need to belong: Desire for interpersonal attachments as a fundamental human motivation. *Psychological Bulletin, 117*, 497–529.
50. Pless, N., & Maak, T. (2004). Building an inclusive diversity culture: Principles, processes and practice. *Journal of Business Ethics, 54*, 129–147.
51. Snyder, C. R., & Fromkin, H. L. (1977). Abnormality as a positive characteristic: The development and validation of a scale measuring need for uniqueness. *Journal of Abnormal Psychology, 86*, 518–527.
52. Demir, M., Simsek, O. F., & Procsal, A. D. (2012). I am so happy 'cause my best friend makes me feel unique: Friendship, personal sense of uniqueness and happiness. *Journal of Happiness Studies, 14*, 1201–1224.
53. Mossakowski, K. N. (2003). Coping with perceived discrimination: Does ethnic identity protect mental health? *Journal of Health and Social Behavior, 44*, 318–331.
54. Mandara, J., Gaylord-Harden, N. K., Richards, M. H., & Ragsdale, B. L. (2009). The effects of changes in racial identity and self-esteem on changes in African American adolescent's mental health. *Child Development, 80*, 1660–1675.
55. Orth, U., Robins, R. W., Widaman, K. F., & Conger, R. D. (2014). Is low self-esteem a risk factor for depression? Findings from a longitudinal study of Mexican-origin youth. *Developmental Psychology, 50*, 622–633.

56. Bhappu, A. D., Zellmer-Bruhn, M., & Anand, V. (2001). The effects of demographic diversity and virtual work environments on knowledge processing in teams. In M. M. Beyerlein, D. A. Johnson, & S. T. Beyerlein (Eds.), *Virtual teams* (pp. 149–165).
57. O'Hara, K. B., Beehr, T. A., & Colarelli, S. M. (1994). Organizational centrality: A third dimension of intraorganizational career movements. *The Journal of Applied Behavioral Science, 30*, 198–216.
58. Guillaume, C., & Pochic, S. (2009). What would you sacrifice? Access to top management and the work–life balance. *Gender, Work and Organization, 16*, 14–36.
59. Dreher, G. F., & Cox, T. H. (2000). Labor market mobility and cash compensation: The moderating effects of race and gender. *Academy of Management Journal, 43*, 890–900.
60. Anier, N., Badea, C., Berthon, M., & Guimond, S. (2018). Perceived acculturation preferences of minority groups and intergroup discrimination: When cultural norms of integration matter. *Journal of Applied Social Psychology, 48*, 506–518.
61. Fletcher, L., & Schofield, K. (2023). Facilitating meaningfulness in the workplace: A field intervention study. *International Journal of Human Resource Management*. (forthcoming).
62. Roberson, Q. M. (2006). Disentangling the meanings of diversity and inclusion in organizations. *Group & Organization Management, 31*, 212–236.
63. Mesmer-Magnus, J. R., DeChurch, L. A., Jimenez-Rodriguez, M., Wildman, J., & Shuffler, M. (2011). A meta-analytic investigation of virtuality and information sharing in teams. *Organizational Behavior and Human Decision Processes, 115*, 214–225.
64. Roberson, Q., & Perry, J. L. (2021). Inclusive leadership in thought and action: A qualitative study of leader perception and behavior. *Group and Organization Management, 47*, 755.
65. Chung, B. G., Ehrhart, K. H., Shore, L. M., Randel, A. E., Dean, M. A., & Kedharnath, U. (2020). Work group inclusion: Test of a scale and model. *Group and Organization Management, 45*, 75–102.
66. O'Leary, M. B., Wilson, J. M., & Metiu, A. (2014). Beyond being there. *MIS Quarterly, 38*, 1219–1244.
67. Mintzberg, H. (1979). *The structuring of organizations: A synthesis of the research*. Prentice-Hall.
68. Theodorakopoulos, N., & Budhwar, P. (2015). Guest editors' introduction: Diversity and inclusion in different work settings: Emerging patterns, challenges, and research agenda. *Human Resource Management, 54*, 177–197.
69. Jackowska, M., & Lauring, J. (2021). What are the effects of working away from the workplace compared to using technology while being at the workplace? Assessing work context and personal context in a global virtual setting. *Journal of International Management, 27*, 100826.
70. Klitmøller, A., & Lauring, J. (2013). When global virtual teams share knowledge: Media richness, cultural distance and language commonality. *Journal of World Business, 48*, 398–406.
71. Olson, J. S., & Olson, G. M. (2006). Bridging distance: Empirical studies of distributed teams. In D. Galletta & P. Zhang (Eds.), *Human computer interaction in management information system* (pp. 23–28). M.E. Sharp Publishers.
72. Rice, D. J., Davidson, B. D., Dannenhoffer, J. F., & Gay, G. K. (2007). Improving the effectiveness of virtual teams by adapting team processes. *Computer Supported Cooperative Work, 16*, 567–594.
73. Maznevski, M. L., & Chudoba, K. M. (2000). Bridging space over time: Global virtual team dynamics and effectiveness. *Organization Science, 11*, 473–492.

Effectiveness of Digital Marketing and Online Customer Behavior in the Hotel Industry

C. Kathiravan and P. Prabu

Abstract While digital marketing has been implemented, the research and development of models and theories guiding the selection and usage of various digital channels are still in their early stages. Digital marketing encompasses a wide range of activities, including but not limited to Internet marketing, e-commerce, mobile marketing, and e-marketing. To gain a deeper understanding of consumers, research on customer behavior focuses on demographics, personality traits, and behavioral patterns such as frequency of use, brand loyalty, advocacy, and propensity to recommend.

This research aims to evaluate the correlation and impact between the effectiveness of digital marketing and online customer behavior in the Chennai hotel industry. The independent variable in this study is digital marketing, while the dependent variables include digital marketing utilization, advantages and disadvantages of digital marketing, and perception of digital marketing. Online customer behavior is a dependent variable, and its sub-variables encompass confidentiality and security, Web site atmosphere, Web site usability, trust, customer service and fulfillment, as well as interactivity/connectivity.

To collect data for this descriptive study, a questionnaire distributed to 305 consumers involved in the Chennai hotel business. The multiple regression analysis was used for data analysis. The results of this study indicates a strong connection and significant influence between the effectiveness of digital marketing and online customer behavior in the Chennai hotel industry.

Keywords Digital marketing · Hotel industry · Effectiveness · Customer behavior

---

C. Kathiravan (✉)
Department of Business Administration, Annamalai University, Deputation to Govt. Arts and Science College Sivakasi, Sivakasi, Tamil Nadu, India

Department of Business Administration, Annamalai University, Govt. Arts and Science College, Sivakasi, India

P. Prabu
Department of Business Administration, Annamalai University, Deputation to Govt. Arts and Science College Sivakasi, Sivakasi, Tamil Nadu, India

J. M. K. P. et al. (eds.), *Business Resilience and Digital Technology in the Post-Pandemic Era*, Contributions to Environmental Sciences & Innovative Business Technology, https://doi.org/10.1007/978-3-031-48075-1_23

1 Introduction

The hotel sector stands as one of the world's oldest and most stable industries, serving as a vital hub for both manufacturing and distribution in today's dynamic business environment. An advantage of hotels is the opportunity for pleasant interactions with customers during their purchases at the front desk. Running a hotel encompasses various costly endeavors, including catering, housekeeping, front desk operations, accounting, marketing, engineering, maintenance and servicing, firefighting, and public relations. Even with the most advanced hotel amenities, the presence of a dedicated staff remains indispensable.

According to Merisavo [1], online marketing encompasses a wide range of channels, such as social media, Web sites, and email. The authors of this study define "digital marketing" as the interaction and communication between businesses or brands and their customers through digital means, including Web sites, email marketing, search engine optimization, and social media marketing platforms like Facebook, WhatsApp, and YouTube. Additionally, content marketing tools like blogs, memes, and info graphics contribute to the digital marketing landscape.

The study of consumer behavior in academia seeks to gain a deeper understanding of why and how individuals purchase, use, and eventually dispose of various products and services. Customer choices are influenced by their morals, ethics, and emotions. Over time, the study of consumer behavior has evolved significantly since its initial exploration in the 1940s and 1950s. It now encompasses a broad range of topics, incorporating elements from various disciplines such as economics, market research, management (specifically behavioral economics), psychology, sociology, anthropology, and ethnology.

2 Review of Literature

In a study by Aden Andrus, it was noted that digital marketing encompasses numerous resources like Web sites and apps, which play a crucial role in advertising and maintaining communication with students. A wide range of digital platforms is employed as part of online marketing strategies, such as emails, SMS, mobile devices, digital TV, virtual videos, blogs, search engine optimization, and social media. The research indicates that the term "digital marketing" encompasses various online channels that can be effectively utilized to attract students and enhance revenue. The study also explores alternative phrases like "electronic marketing" as substitutes for "digital marketing."

According to Kollar and Oberer, digital marketing as the use of digital channels like the Web and mobile devices to market products, services, and ideas. There is a win-win situation when promoting to and educating children about the company. According to the findings, maintaining open lines of communication is crucial to encouraging repeat business.

Hoyer and MacInnis define customer behavior as the study of consumers' activities when exchanging a scarce good for a more plentiful one that better meets their needs. According to Batra and Kazmi [2], a customer's behavior includes their thoughts and feelings before, during, and after utilizing a product or service.

In 1988, Balderjahn [3] conducted a study on consumers who prioritize environmental consciousness. The findings indicated that these eco-conscious individuals are predominantly from higher socioeconomic backgrounds, although distinct predictors exist for each specific behavior. These insights form the basis for market segmentation strategies and can guide policymakers in their educational initiatives.

In 1993, Alwitt and Berger [4] investigated the correlation between strong personal convictions and consumer behavior. While product-specific opinions do correlate with willingness to buy, global perspectives do not.

Hallin [5] argues that people act environmentally conscious for a variety of reasons, including altruism, leadership, and the expectation that one's activities would aid in the survival of the ecosystem. Consumers' optimistic perspectives on environmental issues were not reflected in their spending. Ethical factors rarely impact consumers' decisions. They won't let the things make them less responsible to the environment. Jayawardhena, Wright, and Masterson [6] explore online consumer purchasing behavior, highlighting its nuances and factors influencing customer behavior in the digital realm. The study delves into the effectiveness of digital marketing strategies within the hotel industry, providing insights into online customer behavior that shape purchasing decisions.

3 Research Methodology

3.1 Research Design

The study's objectives and the significance of the results informed the research strategy used. A descriptive methodology is used for this study. Descriptive research is conducted to tell readers everything there is to know about something or someone. The study's methodology provides context for the present. The study's goal is to examine how digital marketing has fared in the hospitality sector and how customers there behave online. The effects of online marketing and customer behavior are examined as they pertain to the hotel industry in Chennai (Fig. 1).

3.2 Objectives of the Study

- To do DM study in the Chennai city hotel business
- To undertake an OCB study in the Chennai city hotel business
- To explore the various viewpoints on the effects of DM on OCB in the Chennai hotel industry
- To examine the effect of DM on OCB

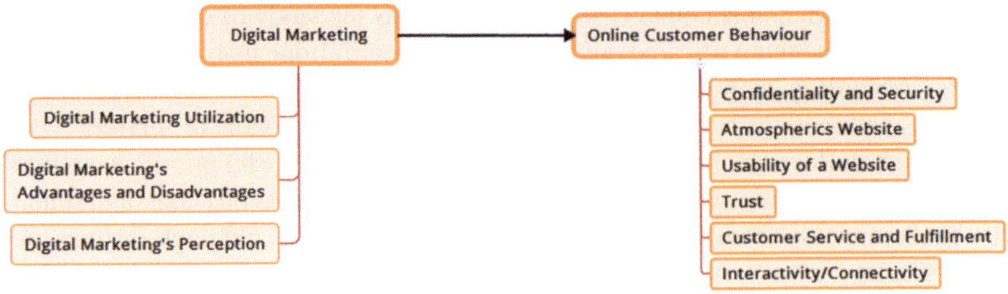

Fig. 1 Framework of the research

3.3 Hypotheses of the Study

- In the Chennai City hotel industry, there is no noticeable change in effect about DM based on demographic profile.
- In the Chennai City hotel industry, there is no obvious variance in OCB by demographic profile.
- In the Chennai City hotel industry, there are no significant debates on the efficacy of DM and OCB.
- There is no evidence that DM affects OCB.

3.4 Data Collection

Visitors to the hotel were the primary source of data, and they provided it by filling out a thoughtful questionnaire. The survey has three sections, the first of which is concerned with the respondents' personal information. Part two focuses on digital marketing, while part three analyzes consumers' online shopping behaviors.

3.5 Reliability Analysis

The reliability of the aforementioned criteria was also determined using reliability analysis. A Cronbach's alpha coefficient of 0.70 or higher indicates satisfactory internal consistency.

S. no.	Variable	Item	Cronbach's alpha
I	Digital marketing	32	0.82
II	Online customer behavior	39	0.85

3.6 Sampling Technique

As was previously said, stratified random sampling is the most appropriate probability sampling strategy for the investigation. Stratified random sampling is a type of probability sampling in which the population is divided into non-overlapping subgroups, or strata, according to variables like geography, gender, or educational attainment.

After the population is stratified, a representative sample is drawn at random from each subset. Each stratum's sample size is calculated according to its relative representation of the whole population.

Stratified random sampling's benefits include accurate representation of the population and a manageable sample size. The results of the study can be more broadly used when using this method because it guarantees that the sample is representative of the total population.

In one scenario, we categorized the healthcare workforce into several groups based on their location and whether or not they were hotel guests. The investigation calls for a probability sampling strategy, and stratified random sampling is the best option.

3.7 Sample Size

In total, 330 surveys were sent out. We got 317 completed surveys. Thirteen surveys were never received. In total, 317 questionnaires were received; 305 were valid responses, leaving 12 invalid ones. That's why there were 305 samples in total.

3.8 Data Analysis

Path analyses can be used to identify potential influencing variables. In this study, the independent variable of "digital marketing" played a significant role. The online activities of consumers are considered a dependent variable.

4 Research Methodology

A random selection of individuals from the construction industry was chosen to complete a questionnaire, resulting in usable data from 305 out of 330 clients surveyed. The collected responses were analyzed using SmartPLS, a statistical software package, to assess the reliability and validity of the data.

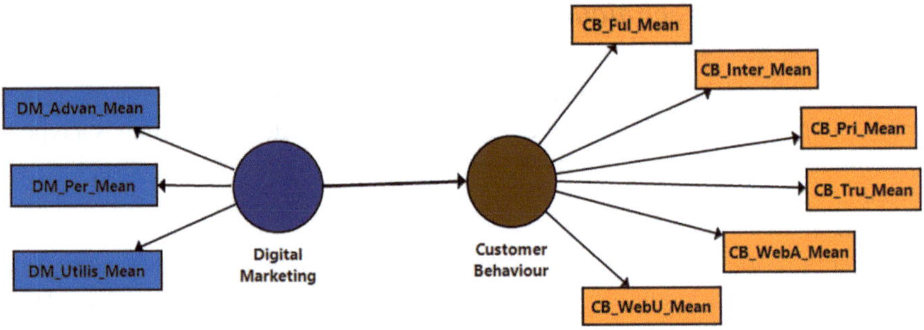

Fig. 2 Structural model

To determine the reliability of the questionnaire, Cronbach's alpha and factor analysis were performed using SPSS. The overall coefficient value of 0.88 indicated a high level of reliability, suggesting that the questionnaire could be trusted. This conclusion was reached through collaborative efforts among researchers [7].

SmartPLS, another analytical tool, was employed to verify the accuracy of the metric and structural models. The study aimed to investigate the impact of digital marketing advancements and changes in online consumer behavior on customer satisfaction in construction projects. The concept of digital marketing was divided into three distinct aspects: "Utilization," "Advantages and Disadvantages," and "Perception." Customer satisfaction was broken down into seven aspects, including "Privacy and Safety," "Website Atmospheres," "Website Usability," "Trust," "Fulfillment and Customer Support," and "Interactivity/Connectivity."

The PLS analysis utilized principal component analysis [8] to track the development of the model's components. PLS has been highly regarded as a powerful analytical tool for enhancing accuracy by Chin et al. [9]. The examination of the PLS model consisted of two stages. The first stage involved evaluating the measurement model, while the second stage assessed the viability of the structural model. The study focused on examining the relationships between the sub-factors (manifest variables) and factors (latent variables) of the model by assessing the construct measures such as validity and reliability (Fig. 2).

4.1 Results and Discussions

4.1.1 Measurement Model

We conducted tests using SmartPLS 2.0.M3 to assess the quality of our model, following the criteria proposed by Tenenhaus et al. [10]. These criteria involve analyzing the structural model as separate layers: the measurement model, the structural model, and the structural regression equation.

To evaluate the reliability of the sub-factors, we utilized SmartPLS software to assess the factor loadings. We adopted a minimum factor loading of 0.45, based on the recommendation of Comrey [11], to identify sub-factors. Additionally, we considered loadings above 0.50, as suggested by Hulland [12], to determine the reliability of the individual sub-factors.

4.1.2 Reliability

To assess construct reliability and internal consistency, we employed composite reliability rather than Cronbach's alpha in PLS-SEM. Composite reliability provides a more suitable measure, as it avoids the assumption of equal consistency among all indicators made by Cronbach's alpha.

4.1.3 Convergence

Table 1 summarizes our evaluation of convergent validity for the credible constructs. We utilized SmartPLS software, along with the AVE test, composite reliability scores, and Cronbach's alpha [13]. All measurements of internal consistency reliability met or surpassed the threshold of 0.7, as recommended by Litwin [14]. This indicates that the dimension sub-factor appropriately represents the respective constructs.

4.1.4 Structural Model Analysis

We employed a bootstrapping re-sampling approach [15] with 380 samples to assess the level of consequence of the pathways established within our model. Figure 3 of

Table 1 Factor loading for indicators of latent constructs

S. no	Factors and sub-factors	Factor loading	Cronbach's alpha	Composite reliability	AVE
1	*Digital marketing*		0.856	0.915	0.783
	Utilization of digital marketing	0.767			
	Advantages & Disadvantages of digital marketing	0.939			
	Perception of digital marketing	0.939			
2	*Online customer behavior*		0.888	0.914	0.642
	Privacy and safety	0.88			
	Web site atmospherics	0.855			
	Web site usability	0.839			
	Trust	0.686			
	Fulfilment and customer support	0.709			
	Interactivity/connectivity	0.859			

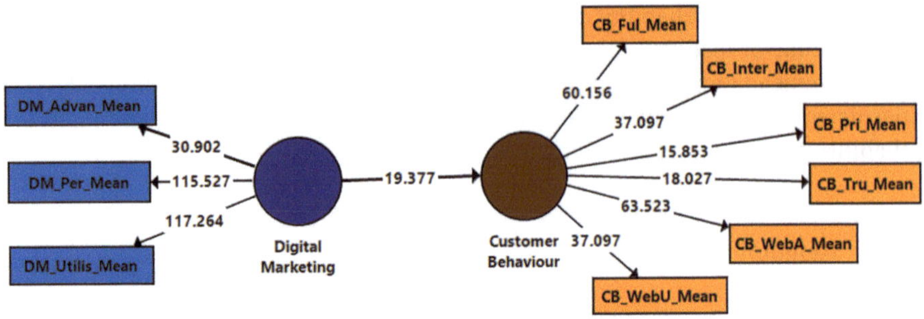

Fig. 3 Bootstrapping diagram

Table 2 Path coefficients along with their bootstrap values and "T" values

Factors	Original sample (O)	Sample mean (M)	Standard deviation (STDEV)	T Statistics	Supported	Significance values
Digital marketing –> customer behavior	0.638	0.640	0.032	20.027	Yes	$p < 0.01$ 0.000**

the structural model, which was validated through our research, illustrates this analysis. The SmartPLS tool was utilized for the examination (Table 2).

- Supporting and significant findings from the initial sample indicate that Online Customer Behavior has a direct and positive effect on Digital Marketing ($\beta = 0.638$, $t = 20.027$, $p = 0.01$).

The data presented here shows that there is a 0.638 significant reduction of one unit in OHS digital marketing and Online Customer Behavior.

5 Suggestions

- Our study aimed to explore the extensive impact of contemporary digital marketing communication channels on customers' decision-making processes. We delved into the roles of brand influencers, opinion leaders, and social institutions such as families, friend groups, and sports leagues in shaping consumers' choices and behaviors. In the current digital landscape, it is crucial for customers to have immediate access to pertinent, adaptable, quantifiable, and personalized information. The digital channels served as a medium, enabling the provision of these essential factors in real time.

6 Conclusion

In the realm of digital marketing in the hotel business in Chennai, a noteworthy study has emerged, shedding light on a previously overlooked connection. This study focuses on the impact of online consumer behavior on the effectiveness of digital marketing. Unlike previous research, the current study investigates this connection in a straightforward manner. In a study conducted by Heinonen [16], it was concluded that digital marketing and online customer behavior are significantly and positively related in the hotel industry. The findings revealed that digital marketing can directly influence online customer behavior, resulting in increased engagement, customer satisfaction, and loyalty. Moreover, it was evident that digital channels play a pivotal role in influencing customers' decision-making processes within the hotel industry.

The practical implications of these findings are significant for hotel industry managers and marketers. They can effectively utilize digital marketing strategies to target and engage with online customers. It is crucial for hotel managers and marketers to adopt a customer-centric approach and consistently monitor and analyze online customer behavior in order to maintain competitiveness in the constantly evolving digital landscape.

This study has made a valuable contribution to the existing literature on digital marketing and online customer behavior within the hotel industry. It offers insights into the various factors that influence customers' decision-making processes. Future research in this area could further expand upon these findings by exploring additional factors that impact online customer behavior in the hotel industry and examining the effectiveness of different digital marketing strategies.

References

1. Merisavo, M. (2006). Theoretical models and measurement issues in interactive marketing communications. *Journal of Business Research, 59*(6), 663–665. https://doi.org/10.1016/j.jbusres.2005.08.004
2. Batra, S. K., & Kazmi, S. H. H. (2008). *Consumer behaviour: Text and cases* (Vol. 1). Excel Books.
3. Balderjahn, I. (1988). Personality variables and environmental attitudes as predictors of ecologically responsible consumption patterns. *Journal of Business Research, 17*(1), 51–56.
4. Alwitt, L. F., & Berger, I. E. (1993). *Understanding the link between environmental attitudes and consumer product usage: Measuring the moderating role of attitude strength*. ACRs North American Advances.
5. Hallin, P. O. (1995). Environmental concern and environmental behavior in Foley, a small town in Minnesota. *Environment and Behavior, 27*(4), 558–578.
6. Jayawardhena, C., Wright, L. T., & Masterson, R. (2003). An investigation of online consumer purchasing. *Qualitative Market Research, 6*(1), 58–65.
7. Hair, J. F., Black, W. C., Babin, B. J., Anderson, R. E., & Tatham, R. L. (2010). *Multivariate data analysis* (7th ed.). Pearson Education.

8. Chin, W. W. (1998). The partial least squares approach for structural equation modeling. In G. A. Marcoulides (Ed.), *Modern methods for business research* (pp. 295–336). Lawrence Erlbaum Associates.
9. Chin, W. W., Marcolin, B. L., & Newsted, P. R. (2003). A partial least squares latent variable modeling approach for measuring interaction effects: Results from a Monte Carlo simulation study and an electronic-mail emotion/adoption study. *Information Systems Research, 14*(2), 189–217. https://doi.org/10.1287/isre.14.2.189.16018
10. Tenenhaus, M., Vinzi, V. E., Chatelin, Y.-M., & Lauro, C. (2005). PLS path modeling. *Computational Statistics & Data Analysis, 48*(1), 159–205. https://doi.org/10.1016/j.csda.2004.03.005
11. Comrey, A. L. (1973). *A first course in factor analysis* (2nd ed.). Academic Press.
12. Hulland, J. (1999). Use of partial least squares (PLS) in strategic management research: A review of four recent studies. *Strategic Management Journal, 20*(2), 195–204. https://doi.org/10.1002/(SICI)1097-0266(199902)20:2<195::AID-SMJ13>3.0.CO;2-7
13. Fornell, C., & Larcker, D. F. (1981). Evaluating structural equation models with unobservable variables and measurement error. *Journal of Marketing Research, 18*(1), 39–50. https://doi.org/10.2307/3151312
14. Litwin, M. S. (1995). *How to measure survey reliability and validity*. SAGE Publications.
15. Efron, B., & Tibshirani, R. J. (1993). *An introduction to the bootstrap*. CRC Press.
16. Heinonen, K. (2011). Consumer activity in social media: Managerial approaches to consumers' social media behavior. *Journal of Consumer Behaviour, 10*(6), 356–364.

MIX
Papier aus verantwortungsvollen Quellen
Paper from responsible sources
FSC® C105338

If you have any concerns about our products,
you can contact us on
ProductSafety@springernature.com

In case Publisher is established outside the EU,
the EU authorized representative is:
Springer Nature Customer Service Center GmbH
Europaplatz 3, 69115 Heidelberg, Germany

Printed by Libri Plureos GmbH
in Hamburg, Germany